Do Bees Build It Best?

Area, Volume, and the Pythagorean Theorem

Teacher's Guide

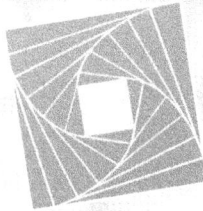

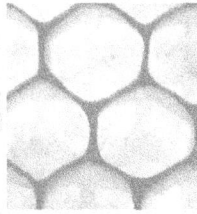

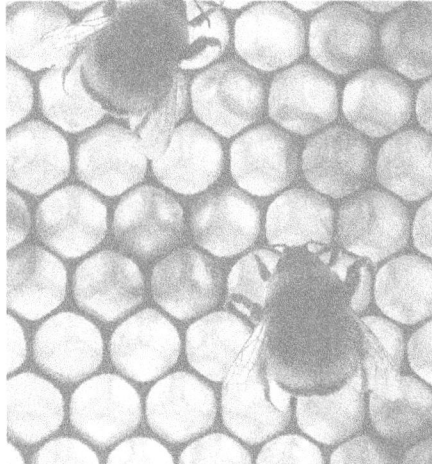

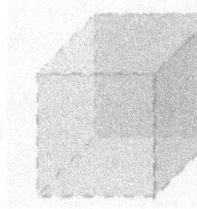

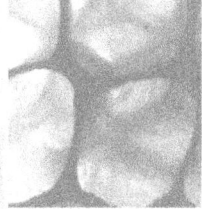

This material is based upon work supported by the National Science Foundation under award numbers ESI-9255262, ESI-0137805, and ESI-0627821. Any opinions, findings, and conclusions or recommendations expressed in this publication are those of the authors and do not necessarily reflect the views of the National Science Foundation.

Key Curriculum
1150 65th Street
Emeryville, California 94608
email: editorial@keypress.com
www.keycurriculum.com

First Edition Authors

Dan Fendel, Diane Resek, Lynne Alper, and Sherry Fraser

Contributors to the Second Edition

Sherry Fraser, Jean Klanica, Brian Lawler, Eric Robinson, Lew Romagnano, Rick Marks, Dan Brutlag, Alan Olds, Mike Bryant, Jeri P. Philbrick, Lori Green, Matt Bremer, Margaret DeArmond

Project Editor

Sharon Taylor

Consulting Editor

Mali Apple

Project Administrator

Juliana Tringali

Professional Reviewer

Rick Marks, Sonoma State University, CA

Calculator Materials Editor

Josephine Noah

Math Checker

Carrie Gongaware

Production Director

Christine Osborne

Production Editor

Andrew Jones

Executive Editor

Josephine Noah

Mathematics Product Manager

Timothy Pope

Publisher

Steven Rasmussen

Contents

Blackline Masters

Calculator Guide and Calculator Notes

Introduction

Do Bees Build It Best? Unit Overview

Intent

In this first unit of Year 2, the honeycombs that bees build serve as the context for developing important ideas in two- and three-dimensional geometry.

Mathematics

The regular form of a honeycomb is striking. Viewed end on, honeycomb cells resemble the hexagonal tiles on a bathroom floor. But a honeycomb is a three-dimensional object, a collection of right hexagonal prisms. Why do bees build their honeycombs this way?

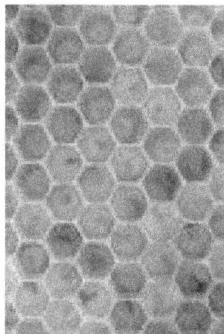

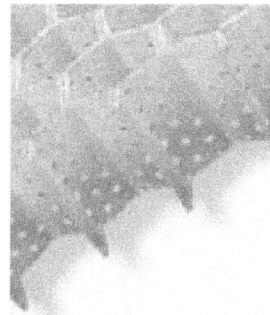

Concepts of measurement—especially area, surface area, and volume—are the mathematical focus of this unit. The main concepts and skills that students will encounter and practice during the unit are summarized by category here.

Area

- Understanding the role of units in measuring area

- Establishing standard units for area, especially those based on units of length

- Recognizing that a figure's perimeter alone does not determine its area

- Discovering formulas for the areas of rectangles, triangles, parallelograms, and trapezoids

- Establishing that a square has the greatest area of all rectangles with a fixed perimeter

- Developing a formula for the area of a regular polygon with a given perimeter in terms of the number of sides

- Discovering that for a fixed perimeter, the more sides a regular polygon has, the greater its area

- Discovering that the ratio of the areas of similar figures is equal to the square of the ratio of their corresponding linear dimensions

The Pythagorean Theorem

- Discovering the Pythagorean theorem by comparing the areas of the squares constructed on the sides of a right triangle
- Proving the Pythagorean theorem using an area argument
- Applying the Pythagorean theorem in a variety of situations

Surface Area and Volume

- Understanding the role of units in measuring surface area and volume
- Establishing standard units for surface area and volume, especially those based on a unit of length
- Recognizing that a solid figure's surface area alone does not determine its volume
- Developing principles relating the volume and surface area of a prism to the area and perimeter of its base
- Discovering that the ratio of the surface areas of similar solids is equal to the square of the ratio of their corresponding linear dimensions, and that the ratio of the volumes of similar solids is equal to the cube of the ratio of their corresponding linear dimensions

Trigonometry

- Reviewing right-triangle trigonometry
- Finding the ranges of the basic trigonometric functions (for acute angles)
- Using the terminology and notation of inverse trigonometric functions

Miscellaneous

- Reviewing similarity
- Reviewing the triangle inequality
- Reviewing the angle sum property for triangles
- Strengthening two- and three-dimensional spatial visualization skills
- Examining the concept of tessellation and discovering which regular polygons tessellate
- Developing some properties of square-root radicals
- Developing the general concept of an inverse function

Progression

The unit begins with an orientation to the context of the unit. Then students develop methods for finding areas of two-dimensional polygons. They look closely at right triangles, reviewing trigonometry and using these ideas to find lengths, angles, and areas. Next students investigate those polygons that maximize area for a given perimeter. They then shift their perspective from two to three dimensions in order to address the unit problem. In addition to the blackline masters provided, students will need geoboards and linking cubes (or other small cubes) during the unit.

Note: Some teachers like to begin Year 2 with a few days of reviewing concepts from Year 1. To allow for this, the first two activities in Supplemental Activities—*Memories of Year 1* and *More Memories of Year 1*—include new problems that build on ideas from each of the units from Year 1 of the IMP curriculum.

Bees and Containers

Area, Geoboards, and Trigonometry

A Special Property of Right Triangles

The Corral Problem

From Two Dimensions to Three

Back to the Bees

Pacing Guides

50-Minute Pacing Guide (31 days)

Day	Activity	In-Class Time Estimate
1	Bees and Containers	10
	Building the Biggest	25
	Introduce: *The Secret Lives of Bees*	10
	Homework: *What to Put It In?*	5
2	*Building the Biggest* (continued)	10
	Discussion: *What to Put It In?*	10
	Area, Geoboards, and Trigonometry	5
	Nailing Down Area	20
	Homework: *Approximating Area*	5
3	Discussion: *Approximating Area*	20
	Nailing Down Area (continued)	25
	Homework: *How Many Can You Find?*	5
4	*Nailing Down Area* (continued)	20
	Discussion: *How Many Can You Find?*	10
	That's All There Is!	20
	Homework: *An Area Shortcut?*	0
5	Discussion: *An Area Shortcut?*	10
	That's All There Is! (continued)	35
	Homework: *Halving Your Way*	5

6	Discussion: *Halving Your Way*	10
	That's All There Is! (continued)	40
	Homework: *The Ins and Outs of Area*	0
7	Discussion: *The Ins and Outs of Area*	15
	Reference: *Parallelograms and Trapezoids*	10
	That's All There Is! (optional discussion of Part III)	10
	Going into the Gallery	15
	Homework: *Forming Formulas*	0
8	Presentations: *The Secret Lives of Bees*	15
	Discussion: *Forming Formulas*	10
	Going into the Gallery (continued)	10
	Introduce: *POW 1: Just Count the Pegs*	10
	The Standard POW Write-up	0
	Homework: *A Right-Triangle Painting*	5
9	Discussion: *A Right-Triangle Painting*	10
	Reference: *A Trigonometric Summary*	5
	More Gallery Measurements	35
	Homework: *Sailboats and Shadows*	0
10	Discussion: *Sailboats and Shadows*	10
	A Special Property of Right Triangles	5
	Tri-Square Rug Games	35
	Homework: *How Big Is It?*	0
11	Discussion: *How Big Is It?*	5
	Tri-Square Rug Games (continued)	20

	Any Two Sides Work	25
	Homework: *Impossible Rugs*	0
12	Discussion: *Impossible Rugs*	10
	Any Two Sides Work (continued)	30
	Homework: *Make the Lines Count*	10
13	Discussion: *Make the Lines Count*	10
	Proof by Rugs	35
	Homework: *The Power of Pythagoras*	5
14	Discussion: *The Power of Pythagoras*	15
	Presentations: *POW 1: Just Count the Pegs*	15
	Introduce: *POW 2: Tessellation Pictures*	20
	Reference: *Some IMPish Tessellations*	0
	Homework: *Leslie's Fertile Flowers*	0
15	Discussion: *Leslie's Fertile Flowers*	25
	Flowers from Different Sides	25
	The Corral Problem	0
	Homework: *Don't Fence Me In*	0
16	Discussion: *Don't Fence Me In*	15
	Rectangles Are Boring!	30
	Homework: *More Fencing, Bigger Corrals*	5
17	Discussion: *More Fencing, Bigger Corrals*	10
	More Opinions About Corrals	40
	Homework: *Simply Square Roots*	0
18	Discussion: *Simply Square Roots*	15

	Building the Best Fence	35
	Homework: *Falling Bridges*	0
19	Discussion: *Falling Bridges*	10
	Building the Best Fence (continued)	40
	Homework: *Leslie's Floral Angles*	0
20	Discussion: *Leslie's Floral Angles*	25
	Presentations: *POW 2: Tessellation Pictures*	10
	From Two Dimensions to Three	0
	Introduce: *POW 3: Possible Patches*	5
	Homework: *Flat Cubes*	10
21	Discussion: *Flat Cubes*	10
	Flat Boxes	40
	Homework: *Not a Sound*	0
22	Discussion: *Not a Sound*	5
	A Voluminous Task	45
	Homework: *Put Your Fist into It*	0
23	Discussion: *Put Your Fist into It*	10
	The Ins and Outs of Boxes	25
	A Sculpture Garden	15
24	*A Sculpture Garden* (continued)	20
	Reference: *The World of Prisms*	10
	Shedding Light on Prisms	15
	Homework: *Pythagoras and the Box*	5
25	Discussion: *Pythagoras and the Box*	10

	Presentations: *POW 3: Possible Patches*	15
	Shedding Light on Prisms (continued)	25
	Homework: *Back on the Farm*	0
26	Discussion: *Back on the Farm*	15
	Which Holds More?	35
	Homework: *Cereal Box Sizes*	0
27	Discussion: *Cereal Box Sizes*	10
	More About Cereal Boxes	40
	Homework: *A Size Summary*	0
28	Discussion: *A Size Summary*	10
	Back to the Bees	0
	Bees Summary Discussion (in teacher's guide)	15
	A-Tessellating We Go	25
	Homework: *A Portfolio of Formulas*	0
29	Discussion: *A Portfolio of Formulas*	45
	Homework: *"Do Bees Build It Best?" Portfolio*	5
30	*In-Class Assessment*	45
	Homework: *Take-Home Assessment*	5
31	Exam Discussion	30
	Unit Reflection	20

IMP Year 2, Do Bees Build It Best? Unit, Teacher Guide

XV

© 2010 Interactive Mathematics Program

90-minute Pacing Guide (20 days)

Day	Activity	In-Class Time Estimate
1	Bees and Containers	15
	Building the Biggest	35
	Introduce: *The Secret Lives of Bees*	10
	Area, Geoboards, and Trigonometry	5
	Nailing Down Area	15
	Homework: *What to Put It In?* and *Approximating Area*	10
2	Discussion: *What to Put It In?*	10
	Discussion: *Approximating Area*	20
	Nailing Down Area (continued)	50
	An Area Shortcut?	10
	Homework: *How Many Can You Find?*	0
3	Discussion: *How Many Can You Find?*	10
	An Area Shortcut? (continued)	25
	That's All There Is!	55
	Homework: *Halving Your Way*	0
4	Discussion: *Halving Your Way*	10
	That's All There Is! (continued)	40
	The Ins and Outs of Area	40
5	Reference: *Parallelograms and Trapezoids*	5
	Going into the Gallery	25

	Presentations: *The Secret Lives of Bees*	15
	Forming Formulas	30
	Introduce: *POW 1: Just Count the Pegs*	10
	The Standard POW Write-up	0
	Homework: *A Right-Triangle Painting*	5
6	Discussion: *A Right-Triangle Painting*	10
	Reference: *A Trigonometric Summary*	10
	More Gallery Measurements	35
	Sailboats and Shadows	30
	A Special Property of Right Triangles	5
	Homework: *How Big Is It?*	0
7	Discussion: *How Big Is It?*	10
	Tri-Square Rug Games	50
	Any Two Sides Work	30
	Homework: *Impossible Rugs*	0
8	Discussion: *Impossible Rugs*	10
	Any Two Sides Work (continued)	25
	Make the Lines Count	30
	Proof by Rugs	25
	Homework: *The Power of Pythagoras*	0
9	Discussion: *The Power of Pythagoras*	15
	Leslie's Fertile Flowers	50
	Flowers from Different Sides	25
	The Corral Problem	0

IMP Year 2, Do Bees Build It Best? Unit, Teacher Guide

xvii

	Homework: *Don't Fence Me In*	0
10	Discussion: *Don't Fence Me In*	20
	Presentations: *POW 1: Just Count the Pegs*	20
	Introduce: *POW 2: Tessellation Pictures*	20
	Reference: *Some IMPish Tessellations*	0
	Rectangles Are Boring!	25
	Homework: *More Fencing, Bigger Corrals*	5
11	Discussion: *More Fencing, Bigger Corrals*	15
	More Opinions About Corrals	40
	Simply Square Roots	35
	Homework: *Falling Bridges*	0
12	Discussion: *Falling Bridges*	15
	Building the Best Fence	75
	Homework: *Leslie's Floral Angles*	0
13	Discussion: *Leslie's Floral Angles*	20
	From Two Dimensions to Three	0
	Flat Cubes	35
	Flat Boxes	35
	Homework: *Not a Sound*	0
14	Discussion: *Not a Sound*	5
	Presentations: *POW 2: Tessellation Pictures*	10
	Introduce: *POW 3: Possible Patches*	5
	A Voluminous Task	40
	The Ins and Outs of Boxes	30

IMP Year 2, Do Bees Build It Best? Unit, Teacher Guide

xviii

	Homework: *Put Your Fist into It*	0
15	Discussion: *Put Your Fist into It*	10
	A Sculpture Garden	35
	Reference: *The World of Prisms*	20
	Shedding Light on Prisms	20
	Homework: *Pythagoras and the Box*	5
16	Discussion: *Pythagoras and the Box*	10
	Shedding Light on Prisms (continued)	20
	Back on the Farm	25
	Which Holds More?	35
	Homework: *Cereal Box Sizes*	0
17	Discussion: *Cereal Box Sizes*	10
	Presentations: *POW 3: Possible Patches*	15
	More About Cereal Boxes	40
	A Size Summary	25
	Homework: *A Portfolio of Formulas*	0
	Back to the Bees	0
	Bees Summary Discussion (in teacher's guide)	15
	A-Tessellating We Go	25
	Discussion: *A Portfolio of Formulas*	45
	Homework: *"Do Bees Build It Best?" Portfolio*	5
	Discussion: *"Do Bees Build It Best?" Portfolio*	30
	In-Class Assessment	45
	Homework: *Take-Home Assessment*	15

Materials and Supplies

All IMP classrooms should have a set of standard supplies and equipment, and students are expected to have materials available for working at home on assignments and at school for classroom work. Lists of these standard supplies are included in the section "Materials and Supplies for the IMP Classroom" in *A Guide to IMP.* There is also a comprehensive list of materials for all units in Year 2.

Listed here are the supplies needed for this unit. General and activity-specific blackline masters are available for presentations on the overhead projector or for student worksheets. The masters are found in the *Do Bees Build It Best?* Unit Resources under Blackline Masters.

Do Bees Build It Best?

- A piece of honeycomb or a large photograph of one
- Construction paper, preferably about 9" x 12" (about four sheets per group)
- A large bag of puffed rice, dried beans, or similar lightweight material
- A large transparent container you can mark on to designate height
- A surface to catch spillage
- Geoboards and rubber bands
- Overhead geoboard
- Geoboard chart paper or graph chart paper (about one sheet per group)
- Glue sticks
- Index cards, approximately 3" x 5" (two per student)
- Grid paper with a 1-inch scale
- Grid paper with a 1-centimeter scale
- Multilink cubes or other cubes (about 54 per group)
- (Optional) Models of various types of prisms (hand-made models will suffice)
- Pattern blocks
- Overhead pattern blocks
- String or thread (about four feet per student)

- Poster paper
- (Optional) Video about bees, beehives, or other content related to the unit

More About Supplies

- Graph paper is a standard supply for IMP classrooms. Blackline masters of 1-Centimeter Graph Paper, ¼-Inch Graph Paper, and 1-Inch Graph Paper are provided so you can make copies and transparencies. (You'll find links to these masters in "Materials and Supplies for Year 2" of the Year 2 guide and in the Unit Resources of each unit.)

IMP Year 2, Do Bees Build It Best? Unit, Teacher Guide

© 2010 Interactive Mathematics Program

xxi

Assessing Progress

Do Bees Build It Best? concludes with two formal unit assessments. In addition, there are many opportunities for more informal, ongoing assessments throughout the unit. For more information about assessment and grading, including general information about the end-of-unit assessments and how to use them, consult the *Year 2: A Guide to IMP* resource.

End-of-Unit Assessments

This unit concludes with in-class and take-home assessments. The in-class assessment is intentionally short so that time pressures will not affect student performance. Students may use graphing calculators and their notes from previous work when they take the assessments. You can download unit assessments from the *Bees* Unit Resources.

Ongoing Assessment

Assessment is a component in providing the best possible ongoing instructional program for students. Ongoing assessment includes the daily work of determining how well students understand key ideas and what level of achievement they have attained in acquiring key skills.

Students' written and oral work provides many opportunities for teachers to gather this information. Here are some recommendations of written assignments and oral presentations to monitor especially carefully that will offer insight into student progress.

- *How Many Can You Find?:* This assignment will inform you about how well students have understood the basics about the meaning of area.

- *That's All There Is!:* This activity will tell you how comfortable students are with a more open-ended approach to area.

- *More Gallery Measurements:* This activity will provide information on students' grasp of the fundamentals of right-triangle trigonometry.

- *Any Two Sides Work, Make the Lines Count,* and *The Power of Pythagoras:* These assignments will tell you about students' comfort with using the Pythagorean theorem.

- *Leslie's Fertile Flowers:* In this activity, students need to combine ideas about area with use of the Pythagorean theorem, so it will give you a sense of their facility with these concepts.

- *More Fencing, Bigger Corrals:* This activity, which involves how changes in linear dimensions affect area, will help you decide how much work students need on this topic.

- *Not a Sound:* This assignment will give you feedback on students' grasp of the concept of surface area.

Supplemental Activities

Do Bees Build It Best? contains a variety of activities at the end of the student pages that you can use to supplement the regular unit material. These activities fall roughly into two categories.

- **Reinforcements** increase students' understanding of and comfort with concepts, techniques, and methods that are discussed in class and are central to the unit.

- **Extensions** allow students to explore ideas beyond those presented in the unit, including generalizations and abstractions of ideas.

The supplemental activities are presented in the teacher's guide and the student book in the approximate sequence in which you might use them. Listed here are specific recommendations about how each activity might work within the unit. You may wish to use some of these activities, especially the later ones, after the unit is completed.

Memories of Year 1 and ***More Memories of Year 1* (reinforcement)** These two sets of problems build on concepts and ideas from each of the units of the Year 1 IMP curriculum.

Measurement Medley—"Length," "Area," and ***"Volume" (reinforcement)*** These three activities offer students experience with fundamental ideas about measurement. Assign them any time over the course of the unit.

How Many of Each? and ***All Units Are Not Equal* (reinforcement or extension)** Both of these activities, which deal with the effect of changing the unit of measurement on the numeric value of area, challenge students to think more deeply about the role of units in measurement. They can be used any time after the initial discussion of the meaning of area and units early in the unit.

***Geoboard Squares* (extension)** This activity, which poses a question that generalizes the findings from the POW *Checkerboard Squares* from the Year 1 unit *Patterns,* can be used any time after students have gained some comfort working with geoboards, perhaps after the discussion of *Nailing Down Area.* The activity requires students to keep careful track of squares of various sizes. (There is a formula for the general case that is not much more complex to state than the *Checkerboard Squares* formula, although it is much harder to find or explain.)

***More Ways to Halve* (extension)** This activity is a follow up to *Halving Your Way.*

***Isosceles Pythagoras* (reinforcement)** In this activity, students are asked to find a simple proof of the Pythagorean theorem for the special case of an isosceles right triangle. This activity can be used any time after the theorem has been stated.

***More About Pythagorean Rugs* (extension)** This activity involves one of the diagrams from the activity *Proof by Rugs,* although this problem is more algebraic in nature than geometric. If students need a hint, you might suggest they examine some cases, including the case of an isosceles triangle.

Pythagorean Proof (extension) This activity, which can be used at any point following the activity *Proof by Rugs,* outlines another proof of the Pythagorean theorem. This proof requires students to use their understanding of similarity and to do some algebraic manipulation of proportion equations.

Pythagoras by Proportion (extension) This activity outlines yet another proof of the Pythagorean theorem. This proof uses the principle that the areas of similar triangles are proportional to the squares of their sides. Based on this principle, the Pythagorean theorem can be paraphrased as essentially saying that the whole is the sum of its parts.

All About Escher (extension) This activity provides a change of pace, asking students to research the life and work of the artist M. C. Escher, and can be used any time after students are familiar with the idea of a tessellation, perhaps in connection with presentations on *POW 2: Tessellation Pictures.*

The Design of Culture (extension) This research activity, which can be assigned any time after the completion of *POW 2: Tessellation Pictures*, provides an opportunity to add a multicultural dimension to the unit.

Tessellation Variations (reinforcement) This activity is aimed primarily at the visual learner and asks students to make tessellation pictures starting from a shape other than the rectangles used in *POW 2: Tessellation Pictures*. This activity can be used after work on that POW has been completed.

Beyond Pythagoras (extension) This activity outlines a proof of the law of cosines, which generalizes the Pythagorean theorem to nonright triangles. The method is similar to the thinking used in the algebraic representation of the situation in *Leslie's Fertile Flowers.* The activity is challenging in the amount of algebra it requires.

Comparing Sines (extension) This activity outlines a proof of the law of sines. As in *Leslie's Fertile Flowers,* the main technique is to find two different expressions for the altitude of a triangle, but this approach uses trigonometry rather than the Pythagorean theorem. The activity is challenging in the amount of algebra it requires.

Hero and His Formula (extension) In *Leslie's Fertile Flowers,* students found the area of a triangle from the lengths of its sides. This activity asks them to prove a formula that generalizes this result and then to simplify the formula using the semiperimeter.

Toasts of the Round Table (extension) This activity, which can be assigned any time after the discussion of *Building the Best Fence,* involves geometric ideas similar to those in the general corral problem.

What's the Answer Worth? (reinforcement) When students measure side lengths, their answers are always approximations. In this activity, which can be used to reinforce ideas about approximation in *Falling Bridges,* students look at the impact of rounding errors.

***Finding the Best Box* (reinforcement)** Students can solve each example of this traditional optimization problem by trial and error and should be able to see the pattern.

***Another Best Box* (extension)** This activity, which can be assigned after the discussion of *The Ins and Outs of Boxes*, involves volume and surface area and requires some algebraic manipulation to find the height in terms of the side of the base.

***From Polygons to Prisms* (reinforcement)** By the beginning of *Back to the Bees*, students understand that among prisms with a fixed height and lateral surface area and whose bases are regular polygons, the more sides the polygon has, the greater the volume of the prism. This activity gives them an opportunity to develop this principle in a more concrete way than is described in the teacher-led *Bees Summary Discussion*.

***What Else Tessellates?* (extension)** This follow-up activity to *A-Tessellating We Go* broadens the concept of tessellation to nonregular polygons, especially nonregular triangles and quadrilaterals.

Bees and Containers

Intent

The three activities in *Bees and Containers* orient students to the context of bees and honeycombs and the general mathematical ideas that will drive the unit.

Mathematics

By examining and building containers of various sizes, students encounter the notions of **volume** and **surface area**. They will also consider other criteria for designing containers, such as cost of materials, what and how much the containers are meant to hold, ease of construction, and storage efficiency. These explorations will help to formalize the statement of the unit problem, which focuses on the criterion of maximizing the volume of a cell of a honeycomb for a given amount of wax (surface area).

Progression

Students will build containers and survey a variety of containers found at home or in stores. They will also conduct research on a topic related to bees.

Building the Biggest

The Secret Lives of Bees

What to Put It In?

IMP Year 2, Do Bees Build It Best? Unit, Teacher's Guide

1

© 2010 Interactive Mathematics Program

Building the Biggest

Intent

This activity will build students' intuition about volume.

Mathematics

This activity helps students to solidify their understanding of the concept of **volume**. It also serves as a vivid illustration of the principle that **prisms** with the same **surface area** can have markedly different volumes. This should move students to at least suspect that *regular* prisms may be more efficient with respect to volume for a given surface area. (Students will be given the vocabulary to accompany this developing intuition as the unit unfolds, so there is no need to define *prism, regularity, volume,* or *surface area* at this time. In future activities students will also explore calculating volume and surface area, so they are not expected to do so here.)

Progression

Using construction paper, groups compete to build a box that will hold the most and then compare the volumes of their boxes by measuring the height of a box's filler after pouring it into a transparent container. The results, combined with a visual comparison of the boxes, serve as the basis for a discussion that begins to probe the question of what characteristics a shape might have in order to be most efficient with respect to volume for a given surface area.

The construction work will probably finish in one class period. Groups may need to save their products for discussion the following day.

Approximate Time

35 minutes

Classroom Organization

Groups, followed by whole-class discussion

Materials

Construction paper (about four sheets per group) and tape
Large bag of puffed rice or similar material for measuring volume
Large transparent container that can be marked to designate height

Doing the Activity

Read the activity as a class and then discuss these questions: How might we measure which box is the biggest? How will we decide which container holds the most? The winner will be determined by filling each construction-paper box with puffed rice, pouring the rice into a large container, and marking how high the rice reaches. Display the measuring container for this purpose. You might point out that this method doesn't indicate *how much* puffed rice a particular box holds, but it is a way of comparing one container to another.

IMP Year 2, Do Bees Build It Best? Unit, *Teacher's Guide*

© 2010 Interactive Mathematics Program

2

Hand out the construction paper and let the groups begin. Caution students not to build portions of the container out of tape alone. If a group wants to start over or try to improve their results, give the group an additional sheet. Giving groups a different color for each subsequent try will make it clear that only a single piece was used to construct each box.

As groups finish, they can use the agreed-upon procedure to compare how much their containers hold. Point out that the puffed rice can't be piled up above the rim of the container.

After a group has finished building their best box, have them consider other possible shapes.

Discussing and Debriefing the Activity

Line up the finished products in ascending order of the amount they hold and use the display as part of the discussion. If possible, tape or staple this progression of boxes to the wall. This will create a visually powerful illustration of some of the principles explored throughout the unit.

Ask students to discuss their discoveries. **What conclusions did you reach about which shapes hold the most?** Allow them to simply explore various ideas, as this is a very complicated question and students come to the issue of volume with various preconceptions.

Key Questions

How might we measure which box is the biggest?

How will we decide which container holds the most?

What conclusions did you reach about which shapes hold the most?

IMP Year 2, Do Bees Build It Best? Unit, Teacher's Guide

3

The Secret Lives of Bees

Intent

Students research an aspect of the lives of bees, increasing their interest in the unit problem, while improving their communication skills.

Mathematics

This activity reinforces the idea that the mathematics students are studying has connections to their other classes and the real world. Additionally, the research, organization, and communication skills students will practice have a universal value.

Progression

Students research and prepare a report on a topic related to bees. (This may be a nice opportunity to collaborate with your school's English and science departments.) Depending upon the level of access students have to reference materials, you may want to schedule class time in the library or computer center soon after assigning this report. Schedule presentations for about a week later.

Approximate Time

10 minutes for introduction
1–3 hours for independent research and report preparation
15 minutes total for presentations

Classroom Organization

Individuals, followed by whole-class presentations

Doing the Activity

Read the activity as a class. Discuss the ground rules for the research paper. What ground rules do you use in other classes for writing research papers? Bring out the ideas that students must use their own words unless quoting directly from another source, in which case they must fully identify the source. Students' most likely source of information will be the Internet. Some students might use more traditional sources, such as encyclopedias and science textbooks.

Also offer guidance about the level of work you expect. Depending on the time available and the needs of your class, you could ask for a well-organized paper two to five pages in length or tell students they only need to list several interesting facts. If you wish to encourage deeper research, emphasize that the report should not be a general overview of the life of bees but should focus on a single aspect of that topic. Be sensitive to the needs of ESL students; extensive writing may be problematic for them. Whatever criteria you establish, present them clearly so students know what your expectations are. Take care that everyone has sufficient opportunities to conduct research. You might take a class day for students to work in the library or computer center.

You might prompt the class to brainstorm possible topics, such as these.
- The role of the queen bee
- Bee sting allergies
- The life cycle of the worker bee
- Bee farms and gathering honey
- Communication among bees
- Bee dancing

Discussing and Debriefing the Activity

Consider these options for handling presentations.
- Have students share their reports in groups, and ask each group to share one fact they discovered.
- Have students read their group members' reports and choose one to present to the class.
- Ask for volunteers to present their reports.

Key Question

What ground rules do you use in other classes for writing research papers?

IMP Year 2, Do Bees Build It Best? Unit, Teacher's Guide

© 2010 Interactive Mathematics Program

5

What to Put It In?

Intent

In preparation for appreciating the basic principle behind the unit problem, the criterion of maximizing volume with respect to surface area, students consider which criteria are relevant to the design of containers.

Mathematics

Students consider the various containers they encounter in their daily lives and the criteria that might be involved in designing those containers. They also encounter the idea that there are multiple units of measurement for volume.

Progression

This activity is best suited as homework. Students sketch containers and reflect on the variety of shapes, materials, units of volume, and design criteria. The follow-up class discussion of possible design criteria leads to refining the unit problem by specifying that a honeycomb should create the maximum storage capacity, per cell, for a given amount of wax.

Approximate Time

5 minutes for introduction
25 minutes for activity (at home or in class)
10 minutes for discussion

Classroom Organization

Individuals, followed by whole-class discussion

Doing the Activity

Have students read the activity in class. Suggest that they may want to stop at a supermarket or other store on the way home from school for ideas.

Discussing and Debriefing the Activity

Begin the discussion by making a class list of shapes and materials used in the containers students found. What shapes did you find? What units of measurement did you encounter? How did the choice of unit match the product? Try to elicit at least one example each of measurement by volume, by weight, and by number.

For Question 4, compile a list of criteria students think are important in designing a container. Look for an opportunity to bring out the idea of storage efficiency, which is the fundamental criterion in the unit problem, as well as other criteria, such as these.

- Cost of materials
- Strength or durability
- Appropriateness to purpose
- Ease of construction

IMP Year 2, Do Bees Build It Best? Unit, Teacher's Guide

6

© 2010 Interactive Mathematics Program

- Safety
- Environmental concern
- The illusion of greater bulk

Have several students present their responses to Question 5. Encourage them to relate their comments to the class list of criteria.

Refining the Unit Problem

Remind students that in this unit they will examine the issue of how well designed the honeycomb is. You might ask, **Which criteria from our list are relevant in evaluating the design of a honeycomb?** Then explain that the unit will focus on one important mathematical criterion of honeycomb design: **A honeycomb should create the maximum storage capacity, per cell, for a given amount of wax.** Post this statement for reference as students proceed through the unit.

Explain that other criteria will be brought up later in the unit and that certain criteria they think are important may not get considered. For now, in the interest of simplicity, students will look only at this one criterion.

Key Questions

What shapes did you find?

What units of measurement did you encounter? How did the choice of unit match the product?

Which criteria from our list are relevant in evaluating the design of a honeycomb?

IMP Year 2, Do Bees Build It Best? Unit, Teacher's Guide

© 2010 Interactive Mathematics Program

7

Area, Geoboards, and Trigonometry

Intent

The activities in *Area, Geoboards, and Trigonometry* lay a conceptual foundation for the concept of area and review basic ideas of trigonometry first encountered in the Year 1 unit *Shadows*.

Mathematics

The mathematical focus of these activities is area. Given that the area of a square with sides of length 1 is 1 square unit, the area of a figure in the plane is a measure of the number of square units that fill the interior of the figure. Reference to the unit square is the common approach, but any plane figure that can cover a plane figure with no overlaps and no gaps can be a unit of measure for area.

Students explore these notions using geoboards, which allow them to quickly create and measure a variety of plane shapes (rectangles, parallelograms, triangles, and trapezoids). Students use patterns to find formulas—actually functions—that relate areas of basic plane figures to their dimensions, and they come to realize that figures can be decomposed into simpler figures to find their areas.

Progression

As *Area, Geoboards, and Trigonometry* draws to a close, students are reintroduced to the six trigonometric ratios developed in the Year 1 unit *Shadows*. Their understanding of these ratios is based on the important idea of similarity, and in particular that all right triangles with a given acute angle are *similar*. In addition, students will begin work on the first POW of the unit. As this is the first POW of Year 2, one aspect of this work will be to reinforce norms for POW write-ups.

Nailing Down Area

Approximating Area

How Many Can You Find?

That's All There Is!

An Area Shortcut?

Halving Your Way

The Ins and Outs of Area

Reference: Parallelograms and Trapezoids

Going into the Gallery

Forming Formulas

A Right-Triangle Painting

Reference: A Trigonometric Summary

POW 1: Just Count the Pegs

Reference: The Standard POW Write-up

More Gallery Measurements

Sailboats and Shadows

IMP Year 2, Do Bees Build It Best? Unit, Teacher's Guide

9

Nailing Down Area

Intent

Students use geoboards to explore the use of units in measuring area. They come to realize that the area of a right triangle can be viewed as half the area of a rectangle.

Mathematics

Throughout the IMP curriculum, students view area measurement as a process of finding how many units (including fractional parts of units) fit inside a given figure, rather than simply as the application of a formula. Geoboards are an important tool for developing this understanding of area. In this activity, students can find the area of each figure by breaking it down into right triangles and applying the intuitive principle that a right triangle is half a rectangle. They can find the areas of rectangles on the geoboard by simply counting square units. Most students soon turn this counting process into a simple "*length* times *width*" principle, without necessarily recognizing this as a formula for area. No formulas for area are introduced yet.

The activity leads into a discussion of common units of area and of the principle that, in order to be useful, a unit of area must tessellate.

Progression

This activity is introduced with a class discussion of the concepts of area and unit of area. The unit of area for this activity is then defined as the square formed by four adjacent pegs on the geoboard. Students familiarize themselves with the geoboard and these concepts by creating squares and right triangles and then finding their areas. The activity asks students to find the area of a number of figures on the geoboard, to find all possible triangles on the geoboard with an area of half a unit, and to find all possible quadrilaterals with an area of one unit. The follow-up discussion includes an exploration of appropriate choices for the shape of a unit of area, introducing the consideration that for a unit of area to be practical, it must tessellate. Finally, students discuss what units are commonly used for area.

Approximate Time

65 minutes

Classroom Organization

Pairs and whole-class discussion

Materials

Geoboards (1 per pair of students)
Rubber bands
Geoboard Paper (2 sheets per student)
Transparent geoboard or *Overhead Geoboard* (1 transparency)

IMP Year 2, Do Bees Build It Best? Unit, Teacher's Guide

10

Doing the Activity

Tell students that before they look at the three-dimensional concept of volume, they will study the simpler, two-dimensional concept of area. You might begin by asking, **How would you define area?** You might list any phrases students use, such as "amount of space inside" and "amount of paper needed to cover an object."

Another way to get at the meaning of area is to ask, **What kinds of things have area?** This may help students distinguish among volume, area, and length, which are all attributes of size.

Area on the Geoboard

Tell students that today they will begin a series of activities using geoboards with the idea of discovering formulas they can use to find the areas of various figures. Introduce the geoboards and let students play constructively for a few minutes to test how to stretch rubber bands around the pegs to form polygons.

Bring out the fact that a unit used for measuring area must itself represent area. Explain that for this activity the unit of area is the small square formed by four adjacent pegs of the geoboard. If necessary, clarify that the pegs themselves are to be treated as points, as if they take up no area.

Ask students to work in pairs to make some simple figures on the geoboard, perhaps starting with rectangles or right triangles, and to find their areas in numbers of unit squares.

As you circulate, get a sense of each student's level of understanding. Students should recognize that the area of a figure is the number of squares (or other area units) that fit inside it. From their work with area in connection with the distributive property and other contexts, they may already have a good grasp of the principle that for rectangles the number of squares can be found by multiplying the number of squares in a row by the number of rows.

Even if students have a good understanding of the formula for the area of a rectangle, do not expect them to use a formula for the area of a triangle yet. Rather, use this opportunity to help them recognize that a right triangle can be viewed as half a rectangle, without formalizing this as $\frac{1}{2}bh$. This formula will be developed gradually over the next several activities.

For instance, students can find the area of the right triangle on the left by thinking of it as half the rectangle on the right. They should notice that the rectangle contains 3 units of area and, as it is composed of two of the triangles, each triangle contains $1\frac{1}{2}$ units of area.

IMP Year 2, Do Bees Build It Best? Unit, Teacher's Guide

© 2010 Interactive Mathematics Program

11

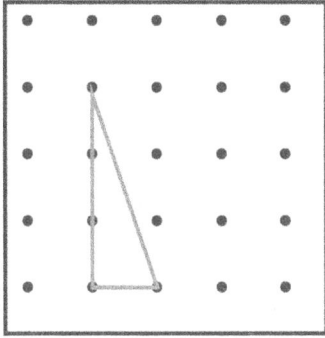

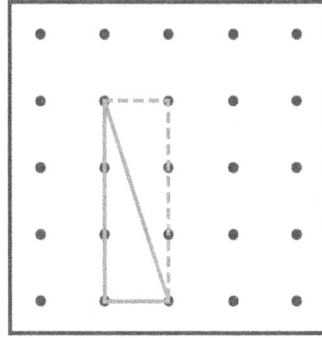

As pairs of students finish finding the areas of their own figures, have them begin work on the activity. Provide each student with two sheets of geoboard paper—one for working through the examples in Question 1 and one for recording the results of Questions 2 and 3.

In Question 2 students may ask whether the mirror image of a triangle or quadrilateral counts as a different shape. However this issue is handled, it will be helpful if everyone counts the same way so students can compare answers. In the Year 1 unit *Shadows*, students were introduced to the concept of congruence and learned that mirror-image figures are considered congruent to each other. You might use this occasion to review that idea.

If students have trouble finding the area of a particular figure, you might suggest surrounding the figure by a large rectangle and subtracting pieces of the rectangle that don't belong. For example, to find the area of the shaded region shown here (figure I in Question 1), they can find the area of the surrounding rectangle by counting squares (or by multiplying length by width) and then subtract the areas of the four right triangles (labeled A, B, C, and D).

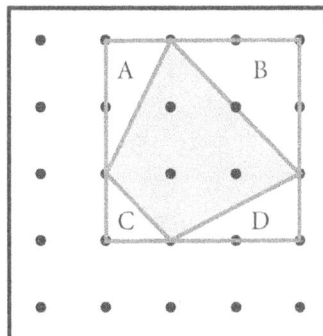

Discussing and Debriefing the Activity

You might choose students at random to give the area of each figure in Question 1. If other students disagree about the answers or have questions, have the presenter explain the reasoning. You might provide a transparent geoboard to assist with the presentations.

For Questions 2 and 3, ask students how many different figures they found and then have someone who found the most figures present his or her work. Encourage students to dispute the areas of figures they question and to contribute new shapes until they can't think of any more.

Area and Units for Area

Review the fact that numeric measurement is intimately tied to the idea of a unit. Point out that students have used two units for area so far: the square on the geoboard and the parallelogram in *Approximating Area* (if you assigned that activity for homework prior to this discussion). (The term **parallelogram** is formally introduced in the reference page *Parallelograms and Trapezoids*, but you may want to use it here.)

Have students brainstorm criteria for selecting a unit for area. One crucial element—that the unit itself must represent area—may be too subtle for students to think to mention, so you may want to bring it out yourself. You might suggest something inappropriate as a unit for area—such as a line segment—and let students explain why it won't work.

Two other elements will be important as students work through the honeycomb problem: the simplicity of the unit and the ability of the units to fit together without overlaps or gaps. Bring out that the square and parallelogram are natural choices for area units because they are simple and can be fitted together to cover the plane. Explain that the ability of these shapes to "cover" or "tile" the plane is expressed by saying that the shapes *tessellate*. You might mention that a POW later in this unit involves creating irregular figures that tessellate.

Ask the class, **What are some units for area that you have encountered before?** Students will likely mention square inches, square feet, square miles, and the like. Bring out that the most common units for area involve squares and that standard units for area are generally squares whose sides are a standard unit of length. There are some exceptions, such as acres.

Key Questions

How would you define area?

What kinds of things have area?

How about using this line segment as a unit for area?

What are some units for area that you have encountered before?

Supplemental Activities

Measurement Medley—"Length," "Area," and "Volume" (reinforcement) offer students experiences with fundamental ideas about measurement. These activities can be used at any time during the unit.

How Many of Each? (reinforcement or extension) and *All Units Are Not Equal* (reinforcement or extension) challenge students to think more deeply about the role of units in measurement. They both deal with the effect of changing the unit of measurement on the numeric value of the area.

Geoboard Squares (extension) poses a question that generalizes the findings from the POW *Checkerboard Squares* from the Year 1 unit *Patterns*. The activity requires students to keep careful track of squares of various sizes. (There is a formula for the general case that is not much more complex to state than the *Checkerboard Squares* formula, although it is much harder to find or explain.)

Approximating Area

Intent

This activity reinforces students' understanding of the basic concept of area.

Mathematics

Having students measure using a nonstandard unit of area reinforces the concept of area. Measuring irregular shapes with an unusual unit of area encourages students to think of area conceptually, rather than as simply a product resulting from the application of a formula.

Progression

Students find the areas of irregular shapes using a given parallelogram as the unit of measure. The follow-up class discussion focuses on what it is that area measures and on the need for a unit of area to tessellate.

Approximate Time

5 minutes for introduction
25 minutes for activity (at home or in class)
20 minutes for discussion

Classroom Organization

Individuals, then groups, followed by whole-class discussion

Doing the Activity

Take a few minutes to introduce the activity. One approach is to draw an arbitrary closed shape (perhaps similar to that in Question 1b) on a grid-paper transparency and ask students to approximate its area using a grid square as the unit of area.

Discussing and Debriefing the Activity

Have students compare answers to Questions 1 and 2 in groups. They can check their answers to Question 2 by trading shapes and measuring.

Bring the class together, and have representatives of various groups report on each question. Reinforce the notion that because area is being measured by covering the diagram with copies of the area unit, the units should not overlap, but should fit together without cracks. (This is known as a **tessellation**.) In other words, the units should cover the surface as completely as possible. Also discuss ways to measure areas with curved boundaries using straight-sided units.

Let one or two students present their results from Question 3, and encourage others to share any additional ideas. Use this discussion to clarify ideas about the meaning of area and to bring out techniques for estimation.

IMP Year 2, Do Bees Build It Best? Unit, Teacher's Guide

15

How Many Can You Find?

Intent

This activity reinforces the concept of area and the notion that the area of a right triangle is half the area of a rectangle.

Mathematics

As students practice finding the areas of polygons, they develop an understanding of determining the area of an irregular polygon by breaking it into smaller pieces. They are again reminded that they can think of the area of a right triangle as being half the area of a rectangle.

Progression

Students work individually on the activity and share ideas and results in a class discussion.

Approximate Time

25 minutes for activity (at home or in class)
10 minutes for discussion

Classroom Organization

Individuals, followed by whole-class discussion

Materials

Geoboard Paper blackline master (2 or more sheets per student)
Transparent geoboard or *Overhead Geoboard* blackline master (1 transparency)

Doing the Activity

This activity requires little or no introduction.

Discussing and Debriefing the Activity

You might choose students at random to give the area of each figure in Question 1. Have the presenter explain his or her reasoning, especially if there is disagreement. You might provide a transparent geoboard to assist with presentations.

For Question 2, ask students how many different shapes they found and have someone present his or her work.

That's All There Is!

Intent

Students discover that triangles with the same base and altitude also have the same area. This discovery will lead students toward an area formula for triangles, which will ultimately come out of their work on *The Ins and Outs of Area*.

Mathematics

This activity provides the foundation for the development of the formula for the area of a triangle by revealing that triangles with the same base and altitude have the same area. Students will ultimately develop that formula in *The Ins and Outs of Area* by looking for patterns relating the base, height, and area of various triangles.

Progression

In Parts I and II, students search for triangles that satisfy certain conditions, and then they look for common properties among their triangles. They prepare a poster that summarizes their work. Part III is reserved for groups that finish early. Students present their findings, reaching the conclusion that triangles with the same base and height have the same area. The terms **altitude** and **base** are introduced in the subsequent discussion, along with the idea that a triangle has three pairs of altitudes and bases. (The case of an altitude falling outside of an obtuse triangle is left for activity *Going into the Gallery*.) An optional discussion of Part III occurs after students work on *The Ins and Outs of Area*, in which they analyze triangles in order to find a formula for area in terms of base and height.

The next several activities build slowly toward the discovery of the area formula. This gradual pace will help students to understand the formula and to feel they could discover it again.

Approximate Time

95 minutes for Parts I and II
10 minutes for optional discussion of Part III (after *The Ins and Outs of Area*)

Classroom Organization

Groups, followed by whole-class discussion

Materials

Geoboards
Rubber bands
Geoboard Paper blackline master (at least 1 sheet per student)
Geoboard chart paper or graph chart paper (1 sheet per group, for presentations)

Doing the Activity

You may want to review the instructions and schedule with students. Part III contains several extensions for groups that finish more quickly.

When groups begin the task of sorting the triangles, let them develop their own schemes for doing so. Some groups will likely come up with the idea of sorting according to the length of the horizontal side. If not, you will need to plant the idea with some groups, because this sorting method will be used as the basis for the principle that the area of a triangle is determined by the length of its horizontal side and its height. (Students will use the more formal terms *base* and *altitude* in later discussion.)

Discussing and Debriefing the Activity

Have groups present their posters, showing how they sorted their triangles.

The subsequent discussion will proceed best if the first groups to present are those who have used "other" methods of sorting. Save for last those groups who sorted their triangles by the length of the horizontal side. When you get to these groups, introduce the term **base** to refer to the horizontal side.

If the class doesn't point out that the triangles in a group all "reach up the same number of rows" on the geoboard, ask, What's the same within each group of triangles, besides the length of the base? Students should realize that the triangles are "equally tall."

Use this opportunity to define an **altitude** of a triangle as the perpendicular line segment from a vertex of the triangle to the line containing the opposite side. Tell students that the length of the altitude is often called the *height* of the triangle. The term *perpendicular* was introduced in the Year 1 unit *Shadows;* you may want to have a student define this term as a review.

Ask students, Is an altitude a side of the triangle? When is it and when isn't it? They should recognize that this is true only in the case of a right triangle. How many altitudes does a triangle have? Bring out that by shifting their point of view, they can use any side of a triangle as the base, so every triangle has three altitudes. (Students will do more work with altitude in the activity *Going into the Gallery*. In particular, they will explore the case of an altitude outside the triangle.)

Return to the posters showing triangles sorted by base length and ask students to use the term *height* (or *altitude*) to describe what the triangles in a given group have in common. They should be able to say something like, Triangles with a given area and the same base also have the same height. Then ask, What does this suggest about what determines the area of a triangle? Elicit the idea that the area of a triangle is somehow determined by its base and height.

Ask, How can the idea that the base and height determine the area be represented using function notation? Try to elicit an equation like $A = f(b, h)$. (Function notation was introduced in the Year 1 unit *The Pit and the Pendulum*.)

Discussion of Part III
Leave the discussion of Part III until after students have completed *The Ins and Outs of Area*, as the requested proofs are most easily constructed using the formula

for the area of a triangle. Depending upon how many groups attempted Part III, and the time available, you may choose to omit this discussion altogether.

For Question 1, probably the best proof that all possible shapes have been found uses the formula for the area of a triangle. The area must be 2, so the product of the base and height must be 4 and the base must be 1, 2, or 4. Because congruent figures are considered the same, students might as well place the base along the bottom of the geoboard. For a base of length 1, there are four possible locations, and students can use any peg in the top row as the location of the third vertex. A similar approach covers the other bases. Students' final task is to eliminate duplicate cases, including cases such as a right triangle with base 1 and height 4 and a right triangle with base 4 and height 1.

For Question 2, there are a total of nine possible areas: $\frac{1}{2}$, 1, $1\frac{1}{2}$, 2, 3, 4, $4\frac{1}{2}$, 6, and 8. Students should be able to recognize the reason for this from the formula for the area of a triangle because the bases can only be of length 1, 2, 3, or 4 and the possible lengths of the altitudes are the same.

The answer to both Questions 3 and 4 is *yes*. Here are two examples of triangles with an area of 2 units.

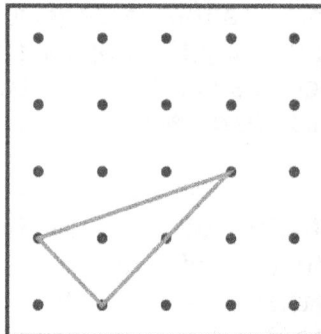

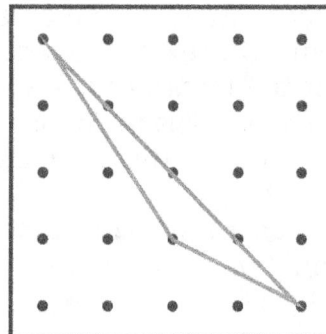

Key Questions

What's the same within each group of triangles, besides the length of the base?

Is an altitude a side of the triangle? When is it and when isn't it?

How many altitudes does a triangle have?

What does this suggest about what determines the area of a triangle?

How can the idea that the base and height determine the area be represented using function notation?

An Area Shortcut?

Intent

This activity reinforces students' developing understanding of area as students discover that figures with the same perimeter may have different areas.

Mathematics

Two important mathematical ideas surface in this activity. First, everyone is asked to find the area of a tracing of her or his hand. As the shape does not lend itself to application of a formula, students are left to estimate the area by counting the enclosed grid-paper squares, a process that emphasizes the basic meaning of area. Second, each student investigates a method for finding area that involves laying string around his or her hand, cutting the string so that its length represents the perimeter, and rearranging it to form a rectangle whose area is easy to find. They will discover that figures with the same perimeter can have different areas.

Progression

Students will work on the two questions individually and share their findings in a class discussion.

Approximate Time

25 minutes for activity (at home or in class)
10 minutes for discussion

Classroom Organization

Individuals, then groups, followed by whole-class discussion

Materials

One-Centimeter Graph Paper blackline master (1 sheet per student; transparency optional)
String or thread (about 4 feet per student)

Doing the Activity

This activity requires little or no introduction.

Discussing and Debriefing the Activity

You might choose one or two students to make transparencies of their work on Question 1. While they prepare to present their results, the rest of the class could try to reach consensus on Question 2 in their groups.

After the presentations of Question 1, have one or two students present their ideas for Question 2. If they perceive nothing wrong with the shortcut, ask groups to examine the areas of different rectangles created by a piece of string or to come up with a counterexample. Students might draw on their geoboard experience to show that figures with the same perimeter can have different areas. The case of 1-by-3 and 2-by-2 rectangles provides a simple illustration of this.

Bring out the contrast between the principle proposed in Question 2 and students' experiences with *Building the Biggest,* in which they realized that three-dimensional figures with the same surface area can have different volumes. Help them to recognize that this is similar to the idea that two-dimensional figures with the same perimeter can have different areas.

IMP Year 2, Do Bees Build It Best? Unit, Teacher's Guide

21

Halving Your Way

Intent
Students practice finding the areas of irregularly shaped polygons.

Mathematics
As students find themselves short of the stated 19 possible ways to accomplish the task of dividing a given rectangle into two parts with equal area, they are forced to employ the organization and categorization skills essential to most exploration. The follow-up discussion brings up the question of proving that there are no further possibilities, perhaps by applying the concept of symmetry.

Progression
Students work on the activity individually. The follow-up discussion moves from verification of their results to the question of how one would prove that all the possibilities had been found.

Approximate Time
5 minutes for introduction
25 minutes for activity (at home or in class)
10 minutes for discussion

Classroom Organization
Individuals, followed by whole-class discussion

Materials
Halving Your Way blackline master (1 per student) or *Geoboard Paper* blackline master (2 sheets per student)

Doing the Activity
You may want to review the rules for this activity with the class.

Discussing and Debriefing the Activity
One way to discuss this activity is to post a sheet of geoboard chart paper and have students each add to the chart one new method for halving the rectangle (perhaps as they come into class, if the activity was done as homework). After everyone has had an opportunity to do this, allow students to post other methods.

Let students comment on the posted methods, examining both whether the diagrams fit the problem conditions and whether they are all unique. When the discussion is finished, have students count the number of distinct methods. If the class doesn't generate all 19 ways, you might leave the problem open, challenging students to continue posting additional methods as they find them.

Whether or not students find all 19 ways, pose the question of how one would prove (without being told how many solutions there are) that a given list is

IMP Year 2, Do Bees Build It Best? Unit, Teacher's Guide

© 2010 Interactive Mathematics Program

22

complete. One related idea you may want to discuss is the role of symmetry. For example, you might suggest that whenever students find a method of halving the rectangle, they can look for other methods that are essentially the same, but reversed.

Key Questions

Do the diagrams fit the conditions? Are they all different?

How would you prove you had all the solutions?

Supplemental Activity

More Ways to Halve (extension) is a follow-up to *Having Your Way*.

IMP Year 2, Do Bees Build It Best? Unit, Teacher's Guide

© 2010 Interactive Mathematics Program

23

The Ins and Outs of Area

Intent

Students analyze data from a number of triangles, searching for a formula for the area of a triangle in terms of base and height, as well as for a geometric explanation of the formula.

Mathematics

In *That's All There Is!,* students realized that triangles with the same base and altitude have the same area. In this activity, they tabulate base, height, and area data for a number of triangles and search for a pattern in order to develop the formula $A = \frac{1}{2}bh$.

As students work toward discovery of this formula, they practice the previously learned skills of recognizing a pattern in a table of data and expressing that pattern symbolically. Question 2 sets the stage for a rich discussion covering proof of the formula and the necessity of considering all possible cases in order to have a complete proof. One possible proof employs the **distributive property** and exposes students to a clear connection between algebra and geometry.

Progression

Students work on the activity individually, share findings in groups, and then participate in a class discussion. The discussion opens with stating and posting the formula for later reference and with a review of how the formula is derived from the information in the In-Out table. Students can then be encouraged to give geometric explanations for the formula (Question 2). These explanations need to consider both right triangles and acute triangles: detailed discussion of oblique triangles can be postponed until after *Going into the Gallery.* Depending upon the nature of student responses and available time, you may wish to present an algebraic approach to the proof, which will make use of the distributive property. This might also be a good time to review notation for polygons, lines, segments, and rays.

Approximate Time

30 minutes for activity (at home or in class)
15 minutes for discussion

Classroom Organization

Individuals, then groups, followed by whole-class discussion

Materials

Geoboard Paper blackline master (1 sheet per student)

Doing the Activity

This activity requires little or no introduction.

IMP Year 2, Do Bees Build It Best? Unit, Teacher's Guide

© 2010 Interactive Mathematics Program

24

Discussing and Debriefing the Activity

Have students share their ideas for Question 1c in their groups. Choose a group to share its In-Out table with the class and state the rule they found. If necessary, have other groups provide additional rows for the table until the rule is clear.

Before discussing explanations for the rule, be sure everyone understands how the pattern emerges from the table. Get a clear statement of the formula, such as, **In any triangle, the area is half the product of the base and the height.** That is,

$$A = \frac{1}{2}bh$$

Confirm the correctness of this formula, and post it with a diagram such as this one.

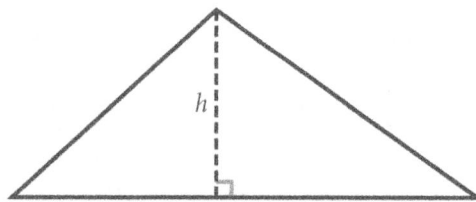

Connect the development of the formula with earlier discussions about function notation by pointing out that this equation gives A as a function of b and h, where

$$f(b, h) = \frac{1}{2}bh$$

Proofs of the Area Formula

Move into a discussion of geometric explanations of the formula, perhaps asking for volunteers to get this started. They might begin with the case of a right triangle, which students looked at informally in *Nailing Down Area*. They should realize that a right triangle is physically half a rectangle with the same base and height and so has half the area of the rectangle. This proves the formula $A = \frac{1}{2}bh$ for right triangles.

For nonright triangles students should recognize that they can't put two copies of the triangle together to make the rectangle, but some students may still be able to explain the relationship geometrically (perhaps with the exception of the case of triangles with an obtuse angle). If necessary suggest that as in the right-triangle case, they can enclose the triangle in a rectangle. Because they know how to deal with right triangles, they can split the triangle into two right triangles. The result is a diagram like this one, in which the shaded figure is the original triangle.

IMP Year 2, Do Bees Build It Best? Unit, Teacher's Guide

25

© 2010 Interactive Mathematics Program

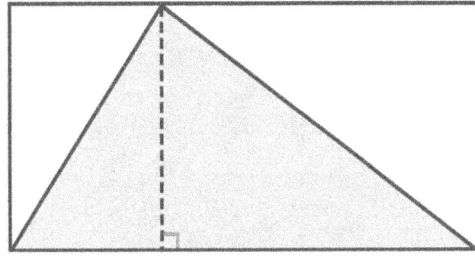

The case in which the altitude to the horizontal side lies outside the triangle will be discussed with *Going into the Gallery*.

Discussed here are two ways—one geometric, one algebraic—to use this diagram to develop the formula for the area of the shaded triangle. Build on what comes most naturally to students and then use your judgment about whether to present both perspectives.

A Geometric Analysis

Labeling the diagram as shown here, ask students what area relationships they notice.

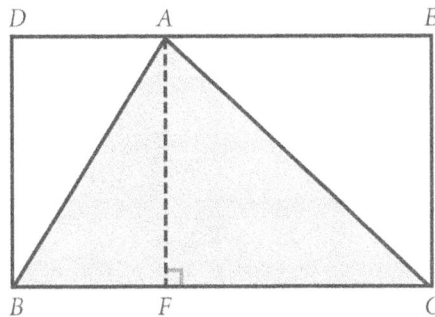

They should be able to articulate these relationships.
- The area of $\triangle AFB$ is half the area of rectangle $AFBD$.
- The area of $\triangle AFC$ is half the area of rectangle $AFCE$.

Students should be able to put these facts together to realize that the area of $\triangle ABC$ is half the area of rectangle $BDEC$. Because the area of the rectangle is *length* times *width,* the area of the triangle is half this product. Bring out that the base and height of $\triangle ABC$ are the same as the length and width of rectangle $BDEC$.

Using Algebra and the Distributive Law

To analyze the area using the established formula for right triangles and the distributive property, label the diagram as shown below. Be sure students understand that the variables *x, y,* and *h* represent the lengths of line segments.

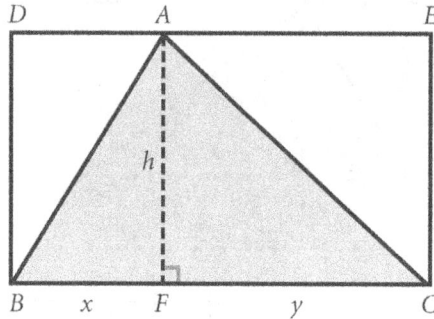

Ask students what they know about area in terms of the variables. They should explain that because $\triangle AFB$ is a right triangle, its area is $\frac{1}{2}xh$; similarly, the area of

$\triangle AFC$ is $\frac{1}{2}yh$. Thus the area of $\triangle ABC$ is $A = \frac{1}{2}xh + \frac{1}{2}yh$.

Reviewing the distributive property as needed, you should be able to get students to rewrite this as $\frac{1}{2}(x+y)h$. They should also realize that $x + y$ is the base b of the

triangle, so the area of $\triangle ABC$ is $\frac{1}{2}bh$.

Reference: Parallelograms and Trapezoids

Intent
These pages define parallelograms and trapezoids.

Mathematics
This reference material defines two special quadrilaterals: **parallelograms,** in which two pairs of sides are parallel, and **trapezoids,** in which exactly one pair of sides is parallel. Students are also reminded of the definitions of quadrilaterals and rectangles and of the naming system for polygons, lines, line segments, and rays. In *Forming Formulas,* students will discover formulas for the areas of parallelograms and trapezoids.

Progression
Although these pages are presented primarily as reference material for students, a brief whole-class overview of the definitions of parallelogram and trapezoid will be useful. Students can then read the more detailed material on their own.

Approximate Time
5 minutes for introduction
10 minutes for individual reading (at home)

Classroom Organization
Individuals

Doing the Activity
Tell students that they will soon be applying their understanding of the areas of rectangles and triangles to develop area formulas for two other types of polygon: parallelograms and trapezoids. Take a few minutes to review the definitions of these two polygons before encouraging students to read this material carefully outside of class.

Discussing and Debriefing the Activity
You may want to ask some follow-up questions, like these, after students have read the information.

Are all rectangles parallelograms?

Are all parallelograms rectangles?

Is a trapezoid also a parallelogram?

What other shapes are also parallelograms?

IMP Year 2, Do Bees Build It Best? Unit, Teacher's Guide

28

Going into the Gallery

Intent

Students think about the altitudes of triangles in a real-world context.

Mathematics

Students deepen their understanding of the concept of **altitude** as the height of a triangle as measured along a perpendicular line from the base to the opposite vertex. The subsequent discussion explores a proof of the area formula for the case of obtuse triangles that also illustrates an application of the **distributive property.**

Progression

Students explore the activity individually or in groups. In the follow-up class discussion, Question 5 addresses the more difficult case of the altitude in an obtuse triangle (which was intentionally omitted from *The Ins and Outs of Area* to avoid confusion). The discussion also affords an opportunity to complete the partial proof of the area formula for triangles that was presented in *The Ins and Outs of Area* by extending it to the obtuse triangle.

Approximate Time

25 minutes

Classroom Organization

Individuals or groups, followed by whole-class discussion

Materials

Going into the Gallery blackline master (1 per student and 1 transparency)
Centimeter rulers or *Metric Ruler and Protractor* blackline master (1 per student)

Doing the Activity

When introducing the activity, you may want to go over the concept of a scale diagram. If you are short on time, you might provide students with the sheet of triangles, which they can freely rotate and mark even if they do not cut the triangles out.

For those who finish early, you might suggest that they find the height of the doorway that would be needed if the triangle in Question 5 were resting on side II or side III and have them prepare a presentation.

Discussing and Debriefing the Activity

Let students present their analyses for the various triangles. Provide copies of the triangles for students to use in their presentations. Students may have made measurements using each side as the horizontal base or may have used alternative approaches to find the lowest possible doorway.

IMP Year 2, Do Bees Build It Best? Unit, Teacher's Guide

© 2010 Interactive Mathematics Program

29

Through the discussion, bring out that the minimum doorway height for a triangular painting is the same as the length of the appropriate altitude of the triangle.

Altitude in Obtuse Triangles

Use this activity to raise the more complicated case of obtuse triangles. Be sure students recognize that in this case, two of the altitudes are outside the triangle. You may need to review the terms *obtuse angle* and *obtuse triangle,* language that will also be important in the discussion of *Tri-Square Rug Games*. For example, ask, What doorway height would be needed if the triangle in Question 5 were placed so that side II was horizontal? If a group worked on this, you might have them make a presentation.

Students should realize that to draw the altitude to side II, they should extend that side and then drop a perpendicular segment from the vertex to this extension, as shown here.

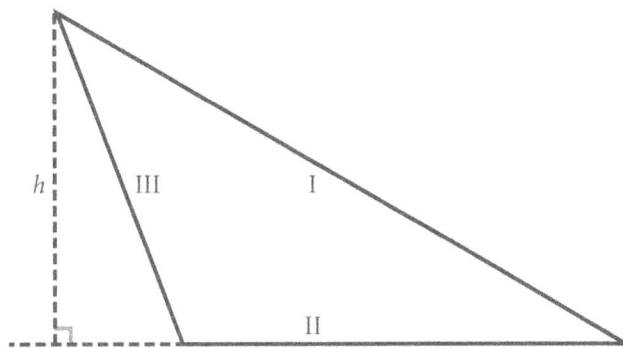

A geoboard may help students understand why the altitude is outside the triangle. The file *Altitudes Bees.gsp* may be useful if you have access to The Geometer's Sketchpad software.

Area of Obtuse Triangles

Question 5 is also a good context in which to discuss the special case of the area formula for obtuse triangles. Start with the diagram from Question 5, with side II horizontal and labeled *b*. Clarify that *b* represents the base of the triangle and not the length of the extended segment.

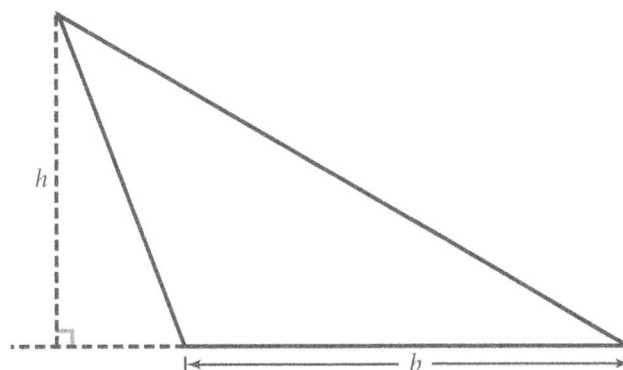

Remind students that the area of a right triangle is easy to understand because two such triangles fit together to make a rectangle, and ask, **What areas can you find in this diagram using right triangles?** If necessary, suggest they label other relevant lengths, as done here.

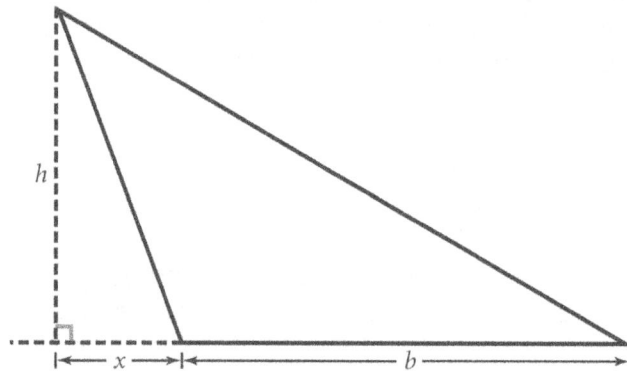

Students should recognize that the diagram has two right triangles.

- A large right triangle, with base $x + b$, height h, and area $\frac{1}{2}(x + b)h$

- A small right triangle, with base x, height h, and area $\frac{1}{2}xh$

They should also realize that the area of the original triangle is the difference between these two areas, or

$$\frac{1}{2}(x + b)h - \frac{1}{2}xh$$

They should be able to manipulate this expression (applying the distributive property) to get $\frac{1}{2}bh$, confirming the formula for the obtuse case.

In the obtuse case (unlike the other cases), the area of the triangle is not half the area of the enclosing rectangle. This diagram shows the obtuse triangle from Question 5 (with side II horizontal) and the rectangle needed to enclose it. Using the earlier notation, the rectangle's area is $(x + b)h$, which is not twice the area of the triangle.

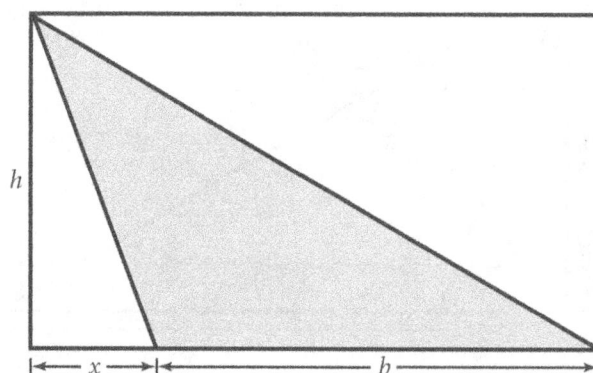

IMP Year 2, Do Bees Build It Best? Unit, Teacher's Guide

31

Key Questions

What does the height of the doorway mean in terms of the triangle?

What areas can you find in this diagram using right triangles?

IMP Year 2, Do Bees Build It Best? Unit, Teacher's Guide

© 2010 Interactive Mathematics Program

32

Forming Formulas

Intent

Students find formulas for the areas of parallelograms and trapezoids.

Mathematics

Students use the strategy of decomposing figures into familiar shapes, finding the areas of those shapes symbolically, and adding the resulting expressions to develop formulas for the areas of parallelograms and trapezoids. By explaining why their formulas work, students come to understand the formulas more deeply and increase the likelihood that they will remember them or be able to reconstruct them in the future.

Progression

Students work individually or in groups to discover area formulas for two special quadrilaterals, with a follow-up class discussion focusing on explanations of the formulas.

Approximate Time

25 minutes for activity (at home or in class)
10 minutes for discussion

Classroom Organization

Individuals or groups, followed by whole-class discussion

Materials

Geoboard Paper blackline master (1 or 2 sheets per student)

Doing the Activity

Introduce the activity. Refer students back to the reference pages *Parallelograms and Trapezoids* for definitions of these polygons, if needed.

If some students already know these formulas, you might emphasize the difference between *knowing* and *understanding* and point out that they need to justify how the formulas work as part of this activity.

You might want to ask volunteers to prepare presentations.

Discussing and Debriefing the Activity

Focus the discussion on students' reasoning about and explanations of the formulas. Here are two common approaches, just two of many possible ways to explain and understand the area formulas.

Rearranging Areas

Cut apart the parallelograms or trapezoids and rearrange them to form figures with the same areas. For example, the shaded area of the parallelogram here has been

IMP Year 2, Do Bees Build It Best? Unit, Teacher's Guide

© 2010 Interactive Mathematics Program

33

moved to form a rectangle with the same base and height as the original parallelogram.

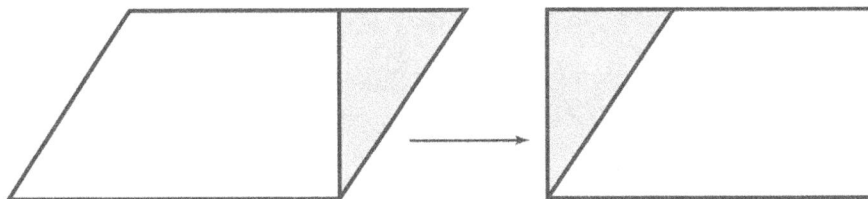

To understand this diagram, students must realize that the slanted side is not the height of the parallelogram, which is shown below with dashed line segment h. Additionally, the base of the figure has not changed.

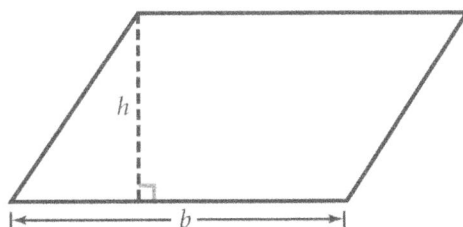

The fact that the parallelogram and rectangle have the same area shows that the area of a parallelogram can be found by multiplying the length of its base by its altitude; that is, the area is the product bh.

Adding Areas of Triangles Using Symbol Manipulation

A more algebraic approach also involves breaking up areas, but instead of actually fitting the pieces together, the areas are combined symbolically. For example, trapezoid $ABCD$ shown here can be thought of as the "sum" of $\triangle ABC$ and $\triangle ACD$. As $\triangle ABC$ has a base of length x and an altitude of length h (the altitude is $\overline{CU}$), the area of $\triangle ABC$ is $\frac{1}{2}xh$. Similarly, the area of $\triangle ACD$ is $\frac{1}{2}yh$ (with altitude $\overline{AV}$). The distributive law can be used to show that the sum of the areas is $\frac{1}{2}(x+y)h$, which is the standard expression for the area of a trapezoid.

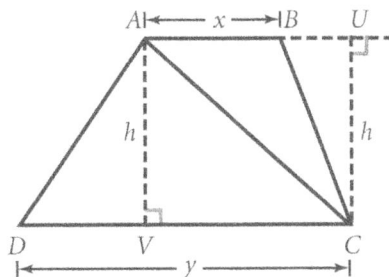

As with the preceding discussion of the parallelogram, a crucial part of analyzing the area of a trapezoid is recognizing that the height of the trapezoid is an important variable.

Key Questions

How might we cut this figure up and rearrange it to form a shape we are more familiar with?

What line segments could we draw that would most easily cut this figure into simpler pieces?

Do we know how to find the areas of these new pieces?

A Right-Triangle Painting

Intent

Students examine ratios within similar right triangles, sparking a review of the basic principles of similarity and right-triangle trigonometry.

Mathematics

Students will use direct measurement to rediscover the principle that among similar right triangles, certain ratios are identical. This flows from the application of a number of concepts introduced in the Year 1 unit *Shadows*, including these.

- The angle sum property for triangles (The sum of the angle measures in any triangle is 180°.)
- The two conditions for similarity (Two triangles are similar if (1) their corresponding angles are equal and (2) their corresponding side lengths are proportional.)
- The uniqueness of triangles in that meeting one of the conditions for similarity dictates that the other has been met
- The right-triangle definitions and application of **sine, cosine,** and **tangent**

Progression

Students work on this activity individually. The discussion of this activity will provide some rich opportunities to review many important topics from Year 1, beginning with noting the differences between students' measurements for the lengths of corresponding sides of the triangles and yet the similarity of the corresponding ratios. It will touch on the topics of similarity, the uniqueness of triangles with regard to similarity due to their rigidity, the angle sum property of triangles, and how similarity ensures equality of the ratios. This will then lead into a brief review of right-triangle trigonometry as presented in the reference pages *A Trigonometric Summary* in the student book.

Approximate Time

5–10 minutes for introduction
15 minutes for independent work
10 minutes for discussion

Classroom Organization

Individuals, then groups, followed by whole-class discussion

Materials

Centimeter rulers and protractors or *Metric Ruler and Protractor* blackline master (1 per student)

Doing the Activity

Students will begin by drawing a 55° right triangle and measuring the length of each side. If this activity is assigned as homework, you may want to accomplish this part in class first, as students might not have protractors and rulers at home.

Emphasize that students need to measure the lengths in Question 1 as carefully as possible to get fairly accurate results for the ratios in Question 2. If their work is sloppy, it won't lead to correct conclusions.

Discussing and Debriefing the Activity

Have students share their numeric results in their groups, and then bring the class together for a discussion of their discoveries. They should have found that even though group members may have drawn triangles of very different sizes, the ratios all came out essentially equal. In other words, they may get different values in Question 1 but should get approximately the same answers for Question 2. (These three ratios correspond to the basic trigonometric relationships, although students are not yet reminded that these ratios are the sine, cosine, and tangent.)

Review the basic ideas of similarity, beginning with a statement of the definition. Two conditions are required for two polygons to be similar.
   •   Corresponding angles must be equal.
   •   Corresponding sides must be proportional.

Ask students, **Why are all the triangles you drew similar?** As needed, remind them that triangles are a special polygon because either one of the two conditions for similarity guarantees the other. In particular, any two triangles with the same set of angles must be similar.

You may want to explore this point further. **Why does one condition for triangles guarantee the other?** You can review the principle of rigidity for triangles—that there is no "play" in an object made by connecting three segments. You might also ask students for an example of two quadrilaterals that have the same set of angles but are not similar, such as a square and a nonsquare rectangle.

How do you know that the activity's triangles all have the same set of angles? Students should recognize that as the triangles are all right triangles, each has a 90° angle, and that each also has a 55° angle, because that was specified.

How do you know that the triangles all have a third angle of 35°? Review the angle sum property briefly if needed.

After it is clearly established that the triangles are all similar, ask, **How does the fact that all the triangles are similar make the ratios equal?** Bring out that this is the other condition of similarity—the proportionality of corresponding sides—and that this proportionality is guaranteed (for triangles) by having corresponding angles equal. Have students explain how this property relates to the ratios in their work on this activity. (*Note:* The ratios in this activity are ratios *within* triangles, while similarity is usually expressed using ratios *between* triangles. Don't let this detail get in the way of the intuitive ideas here. This distinction was the focus of *The Ins and Outs of Proportions* in the Year 1 unit *Shadows*.)

IMP Year 2, Do Bees Build It Best? Unit, Teacher's Guide

© 2010 Interactive Mathematics Program

37

Right-Triangle Trigonometry

Use the material from the reference pages *A Trigonometric Summary* in the student book to review the basic principles of right-triangle trigonometry.

Key Questions

Why did you get approximately the same ratios?

What do you call polygons that have the same shape?

What does similarity mean?

Why are all the triangles you drew similar?

Why do all the triangles you drew have the same set of angles?

How do you know that the triangles all have a third angle of 35°?

How does the fact that all the triangles are similar make the ratios equal?

IMP Year 2, Do Bees Build It Best? Unit, Teacher's Guide

38

Reference: A Trigonometric Summary

Intent

Students are presented with reference material reviewing the right-triangle definitions of sine, cosine, and tangent first developed in the Year 1 unit *Shadows*.

Mathematics

In these reference pages, the **sine**, **cosine**, and **tangent** functions and their reciprocals are defined as ratios of the sides of a right triangle. The development of right-triangle trigonometry is based on the important idea of similar triangles: all right triangles with a specified acute angle are similar, so the six ratios of side lengths will be equal.

Progression

This material is provided primarily as a reminder for students of ideas they encountered in Year 1. The review of the basic trigonometric ratios will flow most naturally from the discussion of *A Right-Triangle Painting*.

Approximate Time

5 minutes

Classroom Organization

Whole class

Doing the Activity

As you conclude the discussion of *A Right-Triangle Painting*, ask students what the ratios found in the activity are called. As needed, review the basic trigonometric ratios with students, directing them to the review of those definitions in the student book.

Depending upon students' familiarity with trigonometry from Year 1, you may need to spend a little extra time on this topic. Be sure students are familiar with such basic right-triangle terminology as *hypotenuse, opposite side*, and *adjacent side*.

IMP Year 2, Do Bees Build It Best? Unit, Teacher's Guide

39

© 2010 Interactive Mathematics Program

POW 1: Just Count the Pegs

Intent

In this first POW of Year 2, students look for formulas that express the area of any polygon on a geoboard as a function of the number of pegs on the perimeter and of the number of pegs in the interior of the polygon.

Mathematics

Students will obtain substantial practice in finding area as they explore the large number of polygons necessary as they collect data in their search for patterns. More importantly, they obtain practice with pattern recognition and are once more reminded of the benefits of collecting sufficient data and of ordering data into some logical sequence in order to facilitate pattern recognition.

The formula relating the area of a geoboard figure to the number of pegs on its border and the number in its interior is called *Pick's theorem,* first published by the Austrian mathematician Georg Pick in 1899.

Progression

The POW is introduced in class, along with a review of what is expected in students' write-ups, followed by presentations about a week later.

Approximate Time

10 minutes for introduction
1–3 hours for activity (at home)
15 minutes for presentations

Classroom Organization

Individuals, followed by whole-class presentations

Materials

Transparent geoboard or *Overhead Geoboard* blackline master (transparency)
Geoboard Paper blackline master (at least 2 sheets per student)

Doing the Activity

Introduce the POW in class to ensure that students understand the three explorations they are asked to complete and that in the final exploration they are looking for a single formula that expresses area as a function of the other two variables. After reading through the POW as a class, illustrate what is meant by "pegs in the interior" and "pegs on the boundary." The "superformula" mentioned in this activity gives the area as a function of these two variables. That is, while Justin and Sarah have each fixed one of the variables and found the area as a function of the other variable, Flashy has taken both variables into account.

Also take a few minutes to review with students *The Standard POW Write-up* in the student book. Remind students that their write-ups require much more than simply giving the formulas they find; they must give their exploration in a clear, orderly manner and thoroughly explain how they went about finding the formulas.

Allow students about a week to work on this POW. On the day before it is due, select three students to make presentations the following day. Give the presenting students transparencies and overhead pens to take home for use in their preparations. You may want to suggest specific parts of the POW for each student to concentrate on. For example, one might present Justin's formula and variations on it, another could focus on Sarah's formula and any variations, and a third could tackle the superformula.

Discussing and Debriefing the Activity

After the three students present their results, ask the rest of the class to contribute any different formulas they may have found. If students did discover the superformula, it is worthwhile to have several students explain how they did so—specifically, how they approached the problem of finding a formula for a function involving two independent variables. For example, they may have conducted systematic explorations of the single-variable cases and noticed a pattern among formulas like Justin's or among formulas like Sarah's. Or they may have combined all the data into a two-variable "master" table with two *Ins—number of pegs on the boundary* and *number of pegs in the interior*—and looked for a pattern there.

If they did not find the superformula, you might leave it as an open question for now, perhaps offering extra credit for a solution.

Whatever level of completeness students achieve on this POW, bring out that their formulas are based on *experimental evidence* rather than on proofs. You might do this by asking, **How sure are you that a new example would fit your formulas?**

Although the patterns in the tables are fairly simple, at least for special cases like Justin's or Sarah's, it is difficult to prove any of the formulas (though it can be done).

Key Question
How sure are you that a new example would fit your formulas?

IMP Year 2, Do Bees Build It Best? Unit, Teacher's Guide

41

© 2010 Interactive Mathematics Program

Reference: The Standard POW Write-up

Intent
Students are given the format and content expectations for Problem of the Week write-ups throughout this course.

Mathematics
Although the mathematical content will vary with each POW, the exercise of writing up the results will play an important role in reinforcing that content. In most cases, the benefits of carefully completing the write-up will be arguably greater than those of mastering the mathematics. The ability to order their thoughts and to present them in a clear, precise manner will deepen and solidify students' understanding of the relevant mathematical principles in a given POW.

Progression
Students who used the IMP Year 1 curriculum will be familiar with what is expected in POW write-ups. Students are introduced to this material, which is presented primarily as reference for their use, in class, and you review the general categories and expectations for POW write-ups.

Approximate Time
10 minutes

Classroom Organization
Individuals

Doing the Activity
Remind students that much more is expected of them in their POW write-ups than merely presenting a solution; each write-up should present—thoroughly, clearly, and in an orderly fashion—the path that led the student to that solution. The reader not only should understand the student's solution, but should come away convinced the solution is correct.

You may also want to take the time to present your general expectations for POWs and review the categories. In particular, you might discuss what role, if any, the self-grading that students do in the self-assessment category will play in the grades you assign.

If students have some experience using rubrics for evaluating written work, or if you wish to introduce this idea, you may want to have them create rubrics to use in assessing POWs. One method is to assign this task to a group at the time the POW is assigned. Emphasize that the rubric the group creates should go beyond general principles—that is, it should be specific to the particular POW.

This will probably work best if you can allow group members some time to work together on both the POW itself and the rubric. If this is not feasible, have

IMP Year 2, Do Bees Build It Best? Unit, Teacher's Guide

42

© 2010 Interactive Mathematics Program

individual group members come up with rubrics and then take some class time to synthesize their work, either with just you and that group or with the whole class.

Ideally, the group will complete its rubric a couple of days before the POW write-ups are due. This will enable you to discuss it with group members and give them time to make needed changes. The group can then present the rubric to the rest of the class before students finish their write-ups. You will then have the option of letting students assess one another's POW write-ups based on that rubric, either instead of or in addition to your own grading.

IMP Year 2, Do Bees Build It Best? Unit, Teacher's Guide

43

More Gallery Measurements

Intent

This activity continues the focus on the altitudes of triangles that began in *Going into the Gallery.* This time the measurements are provided and students use trigonometric functions to find the altitudes.

Mathematics

Students set up and manipulate equations involving the sine, cosine, and tangent in order to find unknown values. As students do this in the context of finding altitudes of triangles, the concept of altitude is reinforced as well.

Progression

Students work on the activity in groups. The follow-up discussion centers on identifying the relevant trigonometric function to use, setting up the equation, and solving that equation to find the altitude.

Approximate Time

35 minutes

Classroom Organization

Groups, followed by whole-class discussion

Doing the Activity

Remind students that when using trigonometric functions, they should always check that their calculators are in degree mode.

You may need to help groups focus on identifying the side they *have* and the side they *want* in terms of the given angle, as these provide clues as to which trigonometric function to use.

Which side of the triangle do we *know?* How would we name that side relative to the angle we are working with? Which side do we *want* to know? What trigonometric ratio uses those two sides?

What equation can we write for this triangle, using the definition of that trig ratio? How can we solve that equation?

For example, in Question 1, students have the length of the hypotenuse and want to find the length of the side opposite the 30° angle. This suggests the sine function and leads to the equation $\sin 30° = \dfrac{h}{94}$.

In Question 3, some students may have decided to find the angle complementary to 61° in order to make this similar to Question 1. Encourage such variations.

IMP Year 2, Do Bees Build It Best? Unit, Teacher's Guide

© 2010 Interactive Mathematics Program

44

Because the altitude in Question 4 is not shown in a vertical position, you might suggest that students turn their books to get a different perspective on the triangle.

You may want to have groups make scale drawings of these triangles based on the given data, measure the altitudes, and compare these measurements to their calculated results.

Discussing and Debriefing the Activity

This is a good opportunity to point out that the degree of precision of the answers should be appropriate to the degree of precision of the information provided.

Key Questions

Which side of the triangle do we *know?* How would we name that side relative to the angle we are working with? Which side do we *want* to know? What trigonometric ratio uses those two sides?

What equation can we write for this triangle, using the definition of that trig ratio? How can we solve that equation?

Sailboats and Shadows

Intent

Students obtain further practice in applying right-triangle trigonometry to real-world situations reminiscent of those encountered in the Year 1 unit *Shadows*.

Mathematics

Students evaluate real-world situations to determine which trigonometric ratio is involved, and they then use trigonometry to solve for an unknown side length in a right triangle.

Progression

Students work on the activity individually or in groups and then briefly share their results in class.

Approximate Time

20 minutes for activity (at home or in class)
10 minutes for discussion

Classroom Organization

Individuals or groups, followed by whole-class discussion

Doing the Activity

Remind students to check that their calculators are in degree mode before they begin.

Discussing and Debriefing the Activity

Have volunteers make presentations on each of the two questions, focusing on the application of the trigonometric functions to solve the problem.

A Special Property of Right Triangles

Intent

The primary aim of these activities is to develop students' understanding of the Pythagorean theorem as a relationship among areas and as a tool for finding lengths. The activities also introduce tessellations.

Mathematics

The **Pythagorean theorem** is one of the most ubiquitous results in geometry. As originally understood in antiquity, the theorem relates the areas of the squares drawn on the sides of a right triangle. The theorem and its converse state that a triangle is a right triangle if and only if the area of the square drawn on the hypotenuse is equal to the sum of the areas of the squares drawn on the two legs.

Using symbols, if a triangle has sides of length *a, b,* and *c,* the triangle is a right triangle if and only if $a^2 + b^2 = c^2$.

In *A Special Property of Right Triangles,* students will explore a proof of the relationship and then use the relationship (1) to find the third side of a right triangle given two side lengths, (2) to determine whether a triangle is a right triangle given three side lengths, and (3) to find the altitudes of a nonright triangle.

Progression

The activities begin with a probability game that introduces the Pythagorean theorem. Students then use the theorem to solve several problems, including finding the altitudes and areas of nonright triangles. In addition, students make presentations on the first of the unit's POWs and begin work on the second.

Tri-Square Rug Games

How Big Is It?

Any Two Sides Work

Impossible Rugs

Make the Lines Count

Proof by Rugs

The Power of Pythagoras

POW 2: Tessellation Pictures

Reference: Some IMPish Tessellations

Leslie's Fertile Flowers

Flowers from Different Sides

IMP Year 2, Do Bees Build It Best? Unit, Teacher's Guide

47

© 2010 Interactive Mathematics Program

Tri-Square Rug Games

Intent

This activity is aimed at getting students to discover the Pythagorean theorem through a game activity like those played in the Year 1 unit *The Game of Pig.*

Mathematics

Tri-square rugs are made of three squares attached at their corners, leaving a triangular space in the center. When a dart randomly hits a rug, one player, Al, wins if the dart hits the largest square. The second player, Betty, wins if the dart hits either of the other squares. This is a fair game if the area of the largest square equals the sum of the areas of the other two squares. Through sorting and observing a variety of such rugs, students discover that when the game is fair—that is, when the area of the largest square equals the sum of the areas of the two smaller squares—the triangular space appears to be a right triangle. Students summarize these observations, which lead to the formulation of the **Pythagorean theorem,** first in terms of areas and then as the famous equation $a^2 + b^2 = c^2$ involving the lengths of the sides. Students also notice that when the area of the largest square is smaller or larger than the sum of the areas of the smaller squares, the triangle is acute or obtuse, respectively.

Progression

Students work in groups, using squares of various sizes to build triangles. They compare the areas of the squares forming the triangles and make posters classifying the triangles based on this area comparison. They report their observations in a class discussion.

Approximate Time

50 minutes

Classroom Organization

Groups, followed by whole-class discussion

Materials

Tri-Square Rug Games blackline master (3 sets per group)
Chart paper (3 sheets per group; alternatively, sheets of construction paper color-coded to the categories "Fair Game," "Al Wins," and "Betty Wins")

Doing the Activity

Because of the open-endedness of this activity, students are sometimes hesitant to begin. Tell such groups to simply start putting tri-square rugs together and figuring out who wins (in the long run) in each case. To do this, they need only compare the area of the largest square with the sum of the areas of the two smaller squares.

Here are some things to suggest or to watch for as you observe groups working.
- Check that students are placing the squares corner to corner appropriately.

IMP Year 2, Do Bees Build It Best? Unit, Teacher's Guide

© 2010 Interactive Mathematics Program

48

- Suggest that students write the area of each square on the figure so that they don't have to compute the area repeatedly.
- Check that students are placing their completed rugs on the appropriate sheet of chart paper. If they fail to do so, the Pythagorean theorem will not be as obvious.
- As groups work on Question 3, suggest that they focus on the triangle formed in the interior of each rug.

Discussing and Debriefing the Activity

Have groups post their charts, grouped by category: "Fair Game," "Al Wins," and "Betty Wins." Let everyone look over the accumulated results, and ask the class to think further about Question 3 in light of all the information. What visual descriptions can you give for how the tri-square rugs are grouped?

Students who think they already understand what's going on can check other groups' results to make sure no examples are out of place. Others may want to revise their analyses based on insights that come from noting this broader range of examples.

Ask two or three students to describe the principle they notice behind the organization of the tri-square rugs. Though individuals may express their ideas differently, their conclusions should be equivalent to these statements.
- All the rugs on the "Fair Game" sheet form right triangles.
- All the rugs on the "Al Wins" sheet form obtuse triangles.
- All the rugs on the "Betty Wins" sheet form acute triangles.

Students may perceive these as tentative conclusions based on experimental evidence. If so, congratulate them for their skepticism but explain that in fact, these conclusions are correct and that they will explore a proof of them shortly.

In discussing these conclusions, you may find the file *Tri-Square Rugs Bees.gsp* useful if you have access to The *Geometer's Sketchpad* software.

In terms of the side lengths of the triangles, the conclusions can be stated symbolically as listed here.

For any triangle with sides of lengths *a, b,* and *c:*
- If $c^2 = a^2 + b^2$, then the angle opposite *c* is a right angle.
- If $c^2 < a^2 + b^2$ (so *c* is "too small"), then the angle opposite *c* is acute.
- If $c^2 > a^2 + b^2$ (so *c* is "too big"), then the angle opposite *c* is obtuse.

The second conclusion holds true even if *c* is not the longest side length.

The Pythagorean Theorem

Tell students that the most important part of their discovery is the connection between right triangles and fair games. Ask, How can you express this discovery about right triangles in terms of areas? Work with students as needed to help them come up with something like this statement.

When a triangle has a right angle, the sum of the areas of the squares on the two shorter sides equals the area of the square on the longest side.

(*Note:* The area analysis in *Tri-Square Rug Games* actually leads more precisely to the converse of the Pythagorean theorem. That is, what students realize is that if $a^2 + b^2$ is equal to c^2, then the triangle formed is a right triangle. This two-directional nature of the theorem will be brought out through specific problems, such as Question 4 of *Any Two Sides Work.*)

Tell students that this principle was known to many ancient civilizations. Though it is unclear who first discovered it, there is evidence of its use in China, India, Egypt, and Babylonia (now Iraq). The principle often played an important role in methods of measurement, especially land surveying. Today this principle is generally called the Pythagorean theorem, after the ancient Greek mathematician and philosopher Pythagoras. Many mathematicians think this is one of the most important theorems in mathematics.

Post the principle in a prominent place, leaving room for the diagram and symbolic summary of it that will be developed next.

You may want to mention that Pythagoras (approximately 580 to 500 BCE.) lived more than two centuries before Euclid. He founded an influential school of thought that made important contributions to mathematics and science and lasted well beyond his death. Copernicus acknowledged that the Pythagoreans had foreshadowed his idea that the earth rotated in an orbit around the sun. None of Pythagoras's writings have survived.

Using Lengths to State the Theorem
Tell students that traditionally we represent the lengths of the legs of a right triangle by the variables *a* and *b* and the length of the hypotenuse by the variable *c,* as shown here.

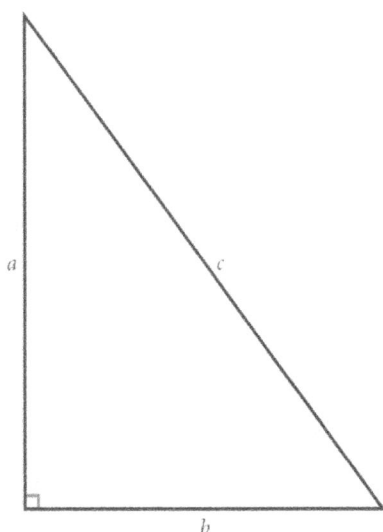

Ask, **How might you express the Pythagorean theorem in terms of the lengths *a, b,* and *c?*** They should be able to come with the equation

$$a^2 + b^2 = c^2$$

to express the area principle. Add the diagram and this equation to the verbal statement of the Pythagorean theorem.

Explain that although students discovered the Pythagorean theorem by thinking about areas, it is most commonly used to find lengths of line segments, especially by considering diagonal lines as hypotenuses of right triangles. Finding the lengths of diagonal lines is key to solving many kinds of mathematics problems.

Key Questions

What visual descriptions can you give for how the tri-square rugs are grouped?

How can you express this discovery about right triangles in terms of areas?

How might you express the Pythagorean theorem in terms of the lengths *a, b,* and *c?*

Supplemental Activity

Isosceles Pythagoras (reinforcement) asks students to prove the Pythagorean theorem for the special case of an isosceles right triangle.

How Big Is It?

Intent

Students synthesize what they have learned about area and measurement and connect their recent work to the unit problem.

Mathematics

Through reflection and summarization, this activity reinforces what has been learned so far: the basic concept of area, the units with which it is commonly measured, and the formulas for calculating the area of a rectangle, triangle, trapezoid, and parallelogram. The activity also asks students to think in more depth about what it means to measure length, area, and volume and to compare the units used for each.

Progression

Students work individually on the activity, which will require little or no discussion.

Approximate Time

20 minutes for activity (at home or in class)
5 minutes for discussion

Classroom Organization

Individuals, followed by a brief group discussion

Doing the Activity

This activity requires little or no introduction.

Discussing and Debriefing the Activity

Have students share their ideas about area and measurement in their groups. You may or may not feel it's worthwhile to discuss the activity as a class.

Key Questions

What key ideas you have learned about area?

How did the class develop those ideas?

IMP Year 2, Do Bees Build It Best? Unit, Teacher's Guide

52

© 2010 Interactive Mathematics Program

Any Two Sides Work

Intent

Students apply the Pythagorean theorem in various contexts, opening the door to a number of important discussion topics.

Mathematics

As they use the Pythagorean theorem, students will come to realize that if they know the lengths of any two sides of a right triangle—the lengths of the two legs or the length of one leg and the hypotenuse—they can find the third length. They are also reminded that a triangle is a right triangle *if and only if* $a^2 + b^2 = c^2$, meaning that if they know the lengths of all three sides, they can also determine whether the triangle is a right triangle.

Question 1 requires students to solve for the hypotenuse; Question 2 requires them to solve for one of the legs. This introduces the manipulation of equations having squares and, ultimately, the concept of *square root,* leading into an introduction to irrational numbers and a discussion of the difference between approximate answers and exact answers in radical form as well as the distinction of a *principal square root.* Questions 3 and 4 require application of the converse of the Pythagorean theorem. Question 4 is an opportunity to illustrate the importance of exact answers because students working with decimal approximations of the square roots will likely come to the wrong conclusion.

Progression

Students work on the activity individually. The follow-up class discussion will cover square roots, irrational numbers, approximations versus exact answers in radical form, and principal square roots.

Approximate Time

55 minutes

Classroom Organization

Individuals, followed by whole-class discussion

Materials

Geoboards and *Geoboard Paper* blackline master (optional)
Rubber bands

Doing the Activity

One way to get students thinking about the situation in Question 1 is to ask how they can be sure the length of the "long diagonal" is not 4 units. The question asks for an estimate and then an exact length. If students interpret "exact" as calling for a several-place decimal rather than a square root, you need not raise the issue now, as this will be clarified in the follow-up discussion.

IMP Year 2, Do Bees Build It Best? Unit, Teacher's Guide

53

© 2010 Interactive Mathematics Program

Discussing and Debriefing the Activity

Question 1

Have presenters include their estimates as part of their discussion of Question 1. At the least they should express that the diagonal is more than 4 units long (the length of each of the legs of the right triangle in which this segment is the hypotenuse) and less than 8 units long (the sum of the legs).

Question 1b is students' first application of the Pythagorean theorem, so encourage them to be especially clear in articulating how they are using the theorem, including explicitly identifying the right triangle to which they are applying the theorem. You can connect this to previous work by asking students to write an equation for the desired length. This should include identifying the variable that represents the length. For example, they might label a diagram like the one below and write the equation $4^2 + 4^2 = d^2$.

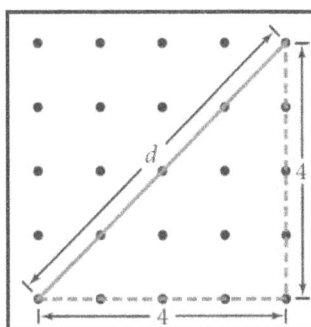

This equation means that $d^2 = 32$, but students may not know what to do with this result. Before going into details, ask, **About how big does this mean d must be?** Elicit the fact that the solution must be a number between 5 and 6. **How does that compare with your estimates?**

Square Roots and Irrational Numbers

Once this very rough estimate of d has been established, lead a more detailed discussion of the concept of a *square root* and the meaning of "exact." You can begin by asking, **What do we call the number whose square is a given number?** As needed, review the term *square root* and the symbol $\sqrt{}$.

Most students will probably be able to get a decimal approximation for $\sqrt{32}$, perhaps one so close that squaring it on the calculator gives a result of exactly 32. This is an important start, and students should be able to find the value of $\sqrt{32}$ to the nearest hundredth (or thousandth, or whatever is appropriate for a given situation) using the calculator. However, also bring out that even a good decimal approximation for $\sqrt{32}$ is not the same as the exact value. You may be able to offer an intuitive argument, based on how decimals are multiplied, for the fact that no finite decimal can have a square exactly equal to 32.

Explain that many square roots are **irrational numbers,** which means they can't be written as fractions and that their decimal expansions go on forever without repeating. In particular, if a whole number is not a perfect square, its square root is irrational.

Tell students that for irrational square roots, we sometimes use an approximation, but to represent the exact value, we have to use the square-root symbol. If they ask why a good approximation isn't enough, mention that rounding off early may produce bigger errors later on. (They will encounter an example of this later, in *Falling Bridges.*)

Two Square Roots
In the context of this problem, the issue of the sign of *d* may not arise naturally, as *d* represents the length of a line segment and therefore must be positive. However, it would be helpful to return to the equation $d^2 = 32$ and ask, **Does the equation $d^2 = 32$ have any other solutions?** If necessary, point out that there is also a negative solution, and ask students how to represent the exact value of this negative solution. They will probably be able to write it as $-\sqrt{32}$.

Explain that the symbol $\sqrt{}$ always represents a nonnegative value—either a positive number or zero—and that this value is called the *principal square root* of the number inside the symbol. When we want to refer to both solutions of the equation $d^2 = 32$, we write $d = \pm\sqrt{32}$, usually read as "plus or minus the square root of 32."

Question 2
After the discussion of Question 1, the work on Question 2 will be fairly straightforward, though students may be surprised at how close the answer is to 8 feet. The answer is $\sqrt{60}$, or approximately 7.75. That is, moving the bottom of the ladder 2 feet from the wall only lowers the top of the ladder by about 3 inches.

Encourage presenters to show which triangle they are using and to write an equation for the missing height, perhaps in the form $a^2 + b^2 = c^2$. For example, they might write $h^2 + 2^2 = 8^2$ and then simplify this to $h^2 = 60$.

Ask, **How is this question different from Question 1? What information did you have in each question?** Bring out that in Question 1 they had the lengths of the legs and were looking for the length of the hypotenuse; in Question 2 they had the lengths of the hypotenuse and one leg and were looking for the length of the other leg.

Question 3
This question touches on the connection between the Pythagorean theorem and its converse. Students will probably get an equation such as $2.5^2 + 1.5^2 = d^2$, which gives $d^2 = 8.5$ and $d \approx 2.9$ meters.

Question 4

This question is more complex than Questions 1 to 3, as it has two stages. If students had trouble with it, you may want to give them another chance to work on it, building on what they have learned from the earlier discussion.

Students may get the lengths of the sides as $\sqrt{2}$, $\sqrt{8}$, and $\sqrt{10}$ or use decimal approximations for the numbers, such as 1.4, 2.8, and 3.2. In either case, ask for an explicit statement of how they will test whether the triangle is a right triangle. **What equation must hold if this is a right triangle?** For example, if they are using approximate values, they should recognize that they are testing a statement like $1.42^2 + 2.82^2 = 3.22^2$. They will probably think that if this statement is true, the triangle is a right triangle, and if it is false, the triangle is not a right triangle.

Based on the approximate values, students should conclude that the statement is false and that the triangle is not a right triangle. Even with somewhat better approximations, they will probably still not get an exact equality. They may decide that the triangle is "almost" a right triangle.

This provides an excellent opportunity to talk about the problems of approximation and to introduce an approach based on exact values represented by the square-root symbol. Again, it is helpful to get students to write down the statement they need to test. They should acknowledge that they need to determine whether this statement is true:

$$\left(\sqrt{2}\right)^2 + \left(\sqrt{8}\right)^2 = \left(\sqrt{10}\right)^2$$

Although this may seem trivial, you will likely have to discuss how to work with expressions like $\left(\sqrt{2}\right)^2$. A question like, **What does "square root" mean?** may be helpful.

Students should then recognize that $\left(\sqrt{2}\right)^2 + \left(\sqrt{8}\right)^2$ is equal to 2 + 8 and $\left(\sqrt{10}\right)^2$ is equal to 10, so the statement is true. The lengths satisfy the equation related to the Pythagorean theorem, so the triangle is a right triangle. Bring out that the use of expressions with square-root symbols makes it clear that the statement is "exactly true" and not just an approximation.

Be on the alert for other methods of proving that the triangle in Question 4 is a right triangle. For instance, both legs are diagonals of squares and form 45° angles with the horizontal. That leaves 90° for the angle between them.

Key Questions

About how big does this mean d must be? How does that compare with your estimates?

What do we call the number whose square is a given number?

Does the equation $d^2 = 32$ have any other solutions?

How is this question different from Question 1? What information did you have in each question?

What equation must hold if this is a right triangle?

Impossible Rugs

Intent
Students review the triangle inequality in the context of tri-square rugs.

Mathematics
Students have been exploring the square of the length of the longest side of a triangle and its relationship to the squares of the lengths of the other two sides. The *triangle inequality* states that the length of the longest side of a triangle must be less than the sum of the lengths of the other two sides, even in a right triangle. Students encountered this principle in connection with the activity *What's Possible?* in the Year 1 unit *Shadows.*

Progression
Working individually, students list combinations of squares that could form tri-square rugs and ones that could not. They use those examples to find a rule to determine whether a given combination will work. This generalization will be a statement of the triangle inequality.

Approximate Time
25 minutes for activity (at home or in class)
10 minutes for discussion

Classroom Organization
Individuals, followed by whole-class discussion

Doing the Activity
This activity requires little or no introduction.

Discussing and Debriefing the Activity
The goal is for students to figure out that for three squares to form a tri-square rug, the length of a side of the largest square must be less than the sum of the lengths of the sides of the other two squares.

If they don't recognize this, ask them to make a tri-square rug using squares of side length 4, 6, and 12, or a similar combination. You may want to use squares on the projector to illustrate the impossibility of doing this.

If there is disagreement about combinations like 4, 5, and 9, point out that the requirement of "a triangular space in the middle" means such a combination is not permitted.

IMP Year 2, Do Bees Build It Best? Unit, Teacher's Guide

58

Ask students to state their conclusions as a principle about triangles, independent of the context of tri-square rugs. They should be able to come up with something like,

The length of the longest side of a triangle must be less than the sum of the lengths of the other two sides.

Identify this principle as the *triangle inequality,* and post it. (Students discovered the triangle inequality in the activity *What's Possible?* in the Year 1 unit *Shadows.*)

IMP Year 2, Do Bees Build It Best? Unit, Teacher's Guide

59

Make the Lines Count

Intent

Students practice applying the Pythagorean theorem to determine the lengths of all possible line segments on the geoboard, expressing the answers exactly and as decimal approximations.

Mathematics

This activity reinforces the Pythagorean theorem and the distinction between exact answers in radical form and decimal approximations. It also gives students another opportunity to determine whether they have found all possible answers to a problem.

Progression

Students work individually on the activity and share findings in groups and a class discussion.

Approximate Time

5–10 minutes for introduction
25 minutes for activity (at home or in class)
10 minutes for discussion

Classroom Organization

Individuals, then groups, followed by whole-class discussion

Materials

Geoboard Paper blackline master (1 sheet per student)

Doing the Activity

Review the example in the activity, clarifying that students are to present answers in two ways: as square roots and as decimal approximations.

Discussing and Debriefing the Activity

Have students come to a consensus in their groups about how many different lengths there are. Poll the groups to get reports on the total. If groups' answers differ, ask for a volunteer to present a list and then let others add lengths that have been omitted or correct any that should not be on the list.

It may be interesting to ask how students developed their lists. Some may have worked geometrically while others simply computed expressions of the form $\sqrt{a^2 + b^2}$, considering all cases with a and b being whole numbers up to 4.

IMP Year 2, Do Bees Build It Best? Unit, Teacher's Guide

60

There are 14 possible lengths. Here is one way to organize them.

$$\sqrt{0^2 + 1^2} = 1 \qquad\qquad\qquad \sqrt{0^2 + 2^2} = 2$$
$$\sqrt{0^2 + 3^2} = 3 \qquad\qquad\qquad \sqrt{0^2 + 4^2} = 4$$

$$\sqrt{1^2 + 1^2} = \sqrt{2} \approx 1.41 \qquad\qquad \sqrt{1^2 + 2^2} = \sqrt{5} \approx 2.24$$
$$\sqrt{1^2 + 3^2} = \sqrt{10} \approx 3.16 \qquad\qquad \sqrt{1^2 + 4^2} = \sqrt{17} \approx 4.12$$

$$\sqrt{2^2 + 2^2} = \sqrt{8} \approx 2.83 \qquad\qquad \sqrt{2^2 + 3^2} = \sqrt{13} \approx 3.61$$
$$\sqrt{2^2 + 4^2} = \sqrt{20} \approx 4.47$$

$$\sqrt{3^2 + 3^2} = \sqrt{18} \approx 4.24 \qquad\qquad \sqrt{3^2 + 4^2} = \sqrt{25} = 5$$

$$\sqrt{4^2 + 4^2} = \sqrt{32} \approx 5.66$$

You may want to remind students that the symbol $\approx$ stands for "is approximately equal to."

Proof by Rugs

Intent

Students explore one well-known proof of the Pythagorean theorem by comparing areas in two diagrams.

Mathematics

Students are asked to determine whether Al and Betty's dart game from the Year 1 unit *The Game of Pig* is fair—that is, whether the shaded areas of the two diagrams are equal. They are then asked how their analysis constitutes a proof of the Pythagorean theorem. This activity reinforces students' knowledge of the theorem, provides an exercise in mathematical proof, and vividly illustrates the pitfall of making assumptions when formulating a proof. In addition, proving that the shaded figure in the second diagram is actually a square requires students to recall that the sum of the angles in a triangle is $180°$.

There are hundreds of proofs of the Pythagorean theorem. The book *The Pythagorean Proposition* by Elisha Scott Loomis (National Council of Teachers of Mathematics, 1968) contains 367 proofs.

Progression

Students work in groups on this activity. Much of the follow-up discussion will focus on the assumption that the shaded area in the second rug is a square and on how to prove that this is the case.

Approximate Time

25 minutes for activity (at home or in class)
10 minutes for discussion

Classroom Organization

Groups, followed by whole-class discussion

Doing the Activity

This activity provides diagrams that students will use to complete a proof of the Pythagorean theorem. Introduce the activity by asking students whether and how they know that the theorem is always true. If necessary, review their work on *Tri-Square Rug Games* and bring out that the fair games appeared to involve right triangles. Emphasize that even if the angles were measured, this would only be approximate evidence for certain cases; proofs must work for all cases and involve more than measurement.

In Question 1, students should reason that the unshaded portions of the two rugs have the same total area because they are composed of the same four triangles.

In Question 2, students might make the assumption that the shaded area in the second rug is a square of side length *c*. As you circulate, ask, *How do you know* that the shaded area in Betty's rug is a square (and not just a rhombus)? Explain that showing that it is a square is an important part of the proof.

Discussing and Debriefing the Activity

Let one or two students explain how they know that the shaded areas of the two rugs are equal. Their explanations will probably involve subtracting the areas of the four triangles from the area of the large square. How they go from this fact to a proof of the Pythagorean theorem will depend on whether they think of the theorem as a statement about area (as originally expressed) or as an equation involving the lengths of the sides.

In the first case, the result from Question 1 is already a proof (except for the issue of why Betty's shaded area is a square). That is, students have shown that the two squares built on the legs of the right triangle have a total area equal to that of the square built on the hypotenuse. This is one formulation of the Pythagorean theorem. (It also happens to be the way the theorem was formulated by Euclid.)

If students think of the Pythagorean theorem as the equation $a^2 + b^2 = c^2$, they will need to express the areas symbolically. In Al's rug, the shaded area is made up of two squares—one of side length a and one of side length b—so the shaded area is $a^2 + b^2$. In Betty's rug, the shaded area is a square of side length c, so the shaded area is c^2. Because the shaded areas are equal, $a^2 + b^2 = c^2$.

Why Is Betty's Shaded Area a Square?

Students may simply assume the shaded area of Betty's rug is a square. If no one offers an explanation, raise the question yourself. Students may initially point out that the four sides have length c, as the sides are all hypotenuses of the original triangle. If this is as far as they get, ask, **Is it possible to have a figure with four equal sides that *isn't* a square?** This should lead them to agree that they need to show that the angles of Betty's shaded figure are all 90°.

There are several ways to prove this. One approach uses this diagram.

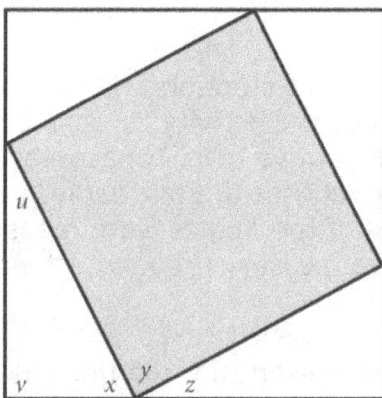

Point out that $x + v + u = 180°$ (because the angles of a triangle add to 180°) and $x + y + z = 180°$ (because these three angles form a straight angle). Because both sums equal 180°, we can set the expressions $x + v + u$ and $x + y + z$ equal to each

IMP Year 2, Do Bees Build It Best? Unit, Teacher's Guide

63

© 2010 Interactive Mathematics Program

other, subtract the x's, and use the fact that $u = z$ to subtract these terms as well. This leaves $v = y$, which shows that y must be 90°. It will probably be helpful to set this up for students as a sequence of equations.

$$x + v + u = x + y + z$$
$$v + u = y + z$$
$$v = y$$

After the class has put together a complete proof of the theorem, you might reinforce their understanding by having several students explain it to the rest of the class. Make sure their proofs include an explanation of why Betty's shaded area is a square.

Key Question

How do you know that the shaded area in Betty's rug is a square?

Is it possible to have a figure with four equal sides that *isn't* a square?

Supplemental Activities

More About Pythagorean Rugs (extension) poses an interesting question about the first of the diagrams from *Proof by Rugs,* although this problem is more algebraic in nature than geometric. If students need a hint, you might suggest they examine some cases, including the case of an isosceles triangle.

Pythagorean Proof (extension) outlines another proof of the Pythagorean theorem. This proof requires students to use their understanding of similarity and to do some algebraic manipulation of proportion equations.

Pythagoras by Proportion (extension) outlines yet another proof of the Pythagorean theorem. This proof uses the principle that the areas of similar triangles are proportional to the squares of their sides. Based on this principle, the Pythagorean theorem can be paraphrased as saying essentially that the whole is the sum of its parts.

IMP Year 2, Do Bees Build It Best? Unit, Teacher's Guide

© 2010 Interactive Mathematics Program

64

The Power of Pythagoras

Intent

Students apply the Pythagorean theorem to real-world situations.

Mathematics

This activity reinforces students' understanding of the Pythagorean theorem as they apply it to several real-world situations. Additionally, Question 3 provides practice with rates.

Progression

Students work individually on the activity and share findings in class. Question 3 will require more discussion than the first two, as the application of the Pythagorean theorem is only part of a larger rate problem.

Approximate Time

30 minutes for activity (at home or in class)
15 minutes for discussion

Classroom Organization

Individuals, followed by whole-class discussion

Doing the Activity

This activity will require little or no introduction.

Discussing and Debriefing the Activity

Questions 1 and 2 are fairly straightforward applications of the Pythagorean theorem, although Question 2 involves several different lengths. If students are having difficulty using the theorem, encourage a lot of detail from presenters.

Ask students to think about whether their answers are reasonable. How could we approximate the answer to this question before doing any calculations? You might point out, for example, that the answer to Question 1 (which is about 16.76 feet) is a little more than the distance from one end of the table to the other.

In both questions, students may run into approximation and rounding issues. Although some may present their answers using square roots, most will probably use decimal values.

In Question 3, the application of the Pythagorean theorem is not the only concern; this problem also involves fundamental ideas about rate, time, and distance. No matter how they approach this question, students will need to find the length of the diagonal in order to determine that Corinne must run 140 meters in the same time that Deanna runs 100 meters.

IMP Year 2, Do Bees Build It Best? Unit, Teacher's Guide

65

Most students will probably first find the length of the diagonal and realize that it takes Deanna 20 seconds to run the diagonal route. They can then figure out that for Corinne to run 140 meters in the same 20 seconds, she has to run 7 meters per second. Some students may notice that they can skip the "time" step by using proportional reasoning. For example, they may set up and solve the equation $\frac{100}{5} = \frac{140}{x}$.

Key Questions

How could we approximate the answer to this question before doing any calculations?

POW 2: Tessellation Pictures

Intent

Students work with geometric designs based on the concept of tessellation.

Mathematics

A **tessellation** is a tiling, or complete covering, of the plane with no overlaps and no gaps. A *regular tessellation* covers the plane with copies of a single regular polygon. There are only three regular tessellations: equilateral triangles, squares, and regular hexagons.

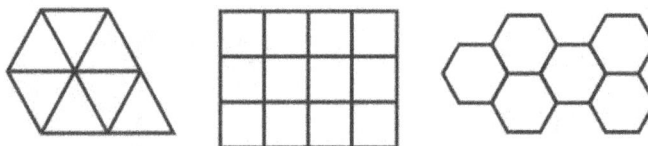

Many other tessellations are possible using other polygons and even curved closed figures, as students will discover in this activity.

Though this POW approaches tessellations more as artistic creations than from a mathematical perspective, the concept of a tessellation plays an important role in resolving the unit problem. That aspect of the problem is treated explicitly in *Back to the Bees.*

Progression

Students work on the POW on their own and share their creations in class.

Approximate Time

20 minutes for introduction
1–3 hours for activity (at home)
10 minutes for discussion and sharing

Classroom Organization

Individuals, followed by whole-class discussion and sharing

Materials

Index cards (several per student)
Poster paper (1 sheet per student)

Doing the Activity

Review the concept of tessellations, which were mentioned briefly in the discussion of *Nailing Down Area.* Use the examples pictured in *Some IMPish Tessellations* in the student book to show how tessellation shapes fit together to cover the plane.

Take a few minutes to go over the instructions for creating a tessellation shape and using it to create a tessellation picture. This can be easily demonstrated by cutting

IMP Year 2, Do Bees Build It Best? Unit, Teacher's Guide

67

© 2010 Interactive Mathematics Program

an index card as described in the POW and laying it on the projector to rearrange the pieces. After taping the rearranged pieces together, turn the shape in various directions and solicit speculation as to what the new shape might represent. With an orientation selected, trace around the shape onto a transparency. Move the shape to each of the four sides of the traced silhouette, tracing it in each position to demonstrate how it tessellates.

The supplemental activity *Tessellation Variations* asks students to make tessellations based on a starting shape other than a rectangle. You may wish to incorporate this idea into the POW. Later in the unit, in *A-Tessellating We Go,* students will consider the question of which regular polygons tessellate.

Discussing and Debriefing the Activity

Have students post their tessellation pictures around the room, and conduct a gallery walk. Then bring the class together to share comments—mathematical or aesthetic—about the work.

Supplemental Activities

All About Escher (extension) gives students the opportunity to explore the art of M. C. Escher, which incorporates many ideas from geometry.

The Design of Culture (extension) is an art-based research activity.

Tessellation Variations (reinforcement) asks students to make tessellation pictures starting from a shape other than a rectangle.

Leslie's Fertile Flowers

Intent

Students apply the Pythagorean theorem to find the altitude and then the area of a nonright triangle.

Mathematics

Students are presented with an acute scalene triangle with known side lengths and are asked to find its altitude and area. They discover that by applying the Pythagorean theorem twice they can find the altitude of a nonright triangle from the lengths of the sides. In the follow-up discussion, students are helped through an algebraic solution that involves writing two equations to represent the geometric situation, simplifying one of the equations through expansion of a binomial, and solving this system of quadratic equations, which becomes linear after some manipulation.

Progression

Students work on the activity individually and share findings in a class discussion. While the activity points them toward a guess-and-check solution, it also presents an opportunity to pursue a rich discussion of an algebraic approach. Although the techniques necessary for this approach are all within the scope of students' algebraic understanding, the complexity of the solution would be a formidable stretch for most students. There is much to be gained by guiding students through this approach, and it is essential that they understand it before tackling the next activity, *Flowers from Different Sides.*

Approximate Time

30 minutes for activity (at home or in class)
25 minutes for discussion

Classroom Organization

Individuals, followed by whole-class discussion

Doing the Activity

If you think some students may struggle with how to begin, point out the guess-and-check approach mentioned in Question 1. The numbers were chosen so that the answer would be fairly easy to guess. The real focus of the activity is on how to be sure the guess is correct.

Discussing and Debriefing the Activity

The important part of this activity is the justification students give for the answer. The key ideas are that the two right triangles formed each have a leg of length h and that the 14-foot side must be split up so that the same value for h fits the Pythagorean theorem when applied in both triangles.

IMP Year 2, Do Bees Build It Best? Unit, Teacher's Guide

69

© 2010 Interactive Mathematics Program

If no one can explain this procedure, you might ask, Suppose the side of length 14 were split into segments of lengths 6 and 8. Why wouldn't that work?

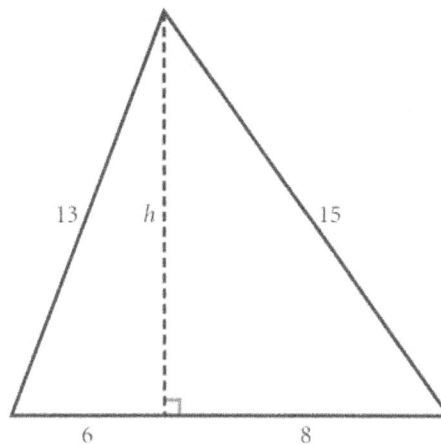

Ask why this diagram is impossible. As a further hint, ask, What would h be in this diagram? You might have half the class calculate h for one of the right triangles and the other half calculate it for the remaining triangle. It may be helpful to split the diagram into two separate triangles.

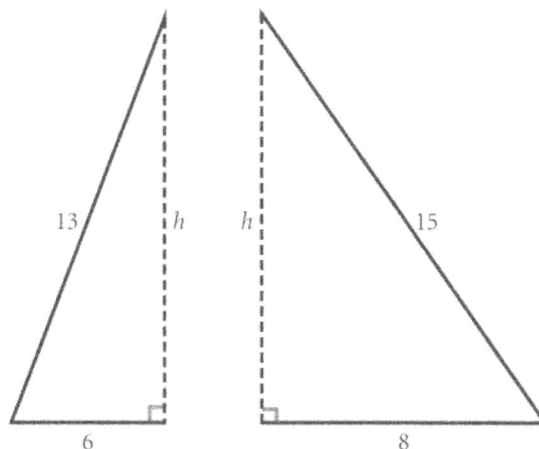

The right triangle on the left gives $h^2 + 6^2 = 13^2$, which simplifies to $h^2 = 133$. The other right triangle gives $h^2 + 8^2 = 15^2$, so $h^2 = 161$. As h must be the same in both cases, splitting the side into segments of lengths 6 and 8 must be incorrect. Students should verify that with segments of lengths 5 and 9, the two right triangles give the same value for h (namely, 12), so this must be the right answer.

Ask a volunteer for the area of the flowerbed in Question 3 ($\frac{1}{2} \cdot 14 \cdot 12 = 84$ square feet). This result will be needed in the next activity, *Flowers from Different Sides.*

Setting Up the Solution Symbolically
This activity gives a nice opportunity to discuss the application of symbols, reviewing both using variables to represent unknowns and equation solving.

Begin by asking the class, **How might you have found the right way to split up the base of the triangle without guessing, or if the numbers weren't whole numbers?** If necessary, suggest they use a variable—say x—to represent one of the segments making up the base. **How would you represent the other segment algebraically?** If they suggest calling it y, ask if there's anything they know about the relationship between x and y. Particularly, ask, **Could you have $x = 6$ and $y = 9$?** They should realize that this is impossible because the two values don't add to 14.

Then ask, **How might we express the other segment in terms of x?** They should know that the other segment is found by subtracting x from 14. If necessary, have them find the other segment for several values of x and perhaps list the results in an In-Out table. The diagram can now be labeled like this.

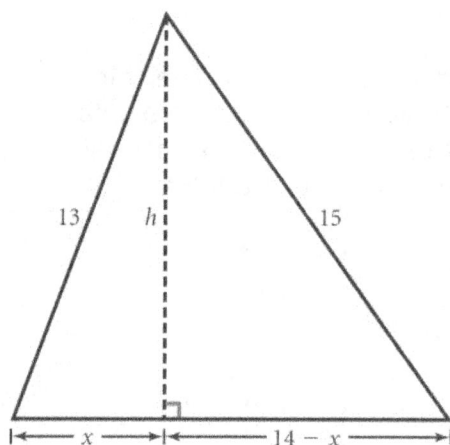

With this introduction, have groups discuss how they would use this diagram to write an equation that would reveal what x is.

If necessary, ask them to restate how they knew in the original problem that the base split into segments of lengths 5 feet and 9 feet, or suggest they use the Pythagorean theorem to write two equations involving both x and h. These questions should help lead the class to the equations $x^2 + h^2 = 13^2$ and $(14 - x)^2 + h^2 = 15^2$.

You will likely need to help the class eliminate h from these equations. One approach is to ask what expression they would get if they solved each equation for h^2. **What does the first equation say about the value of h^2? What about the second equation?** With some assistance, they should eventually find an equation like $13^2 - x^2 = 15^2 - (14 - x)^2$.

Students will return to this equation in the unit *Fireworks* in connection with work on simplifying expressions, including expanding quadratic expressions. For now, you might have them simply use the equation to verify the numeric solution ($x = 5$)

they already found, or let them explore how to solve the equation using the algebra they know. (The x^2 terms cancel out, so this becomes a linear equation.)

Key Questions

Suppose the side of length 14 were split into segments of lengths 6 and 8.

Why wouldn't that work?

What would h be in this diagram?

How might you have found the right way to split up the base of the triangle without guessing, or if the numbers weren't whole numbers?

How would you represent the other segment algebraically?

How did you know that the original base split into segments of lengths 5 and 9?

How can you use the Pythagorean theorem to write equations involving both x and h?

What does the first equation say about the value of h^2? What about the second equation?

Supplemental Activities

Beyond Pythagoras (extension) uses the reasoning just described for this activity to outline a proof of the law of cosines, which generalizes the Pythagorean theorem to nonright triangles. The method is similar to the thinking used in the algebraic representation of the situation in *Leslie's Fertile Flowers.* The activity is challenging in the amount of algebra it requires.

Comparing Sines (extension) outlines a proof of the law of sines. As in *Leslie's Fertile Flowers,* the main technique is to find two different expressions for the altitude of a triangle, but this approach uses trigonometry rather than the Pythagorean theorem. This activity is also challenging in the amount of algebra required.

Hero and His Formula (extension) uses the approach described in this activity as the basis for developing a formula for the area of a triangle in terms of its side lengths.

Flowers from Different Sides

Intent

This activity gives students additional practice with applying the Pythagorean theorem, with the formula for the area of a triangle, and with the idea that any side of a triangle can be the base.

Mathematics

This activity reinforces several concepts: the Pythagorean theorem, the area of a triangle, and the algebraic method just introduced in *Leslie's Fertile Flowers.* More importantly, it reminds students that every triangle has three altitudes and that area is a measurement—not simply the result of a formula—that does not change with a figure's orientation.

Progression

Students work individually or in groups to find the area of the triangle from *Leslie's Fertile Flowers* from the perspective of the remaining two altitudes. The follow-up class discussion explores the ideas that if the area of a triangle is known, the formula for the area can be solved for any of the altitudes; and that area is a measurement that is not affected by a figure's orientation.

Approximate Time

15 minutes for activity
10 minutes for discussion

Classroom Organization

Individuals or groups, followed by whole-class discussion

Doing the Activity

When introducing the activity, emphasize the fact that every triangle has three altitudes.

Discussing and Debriefing the Activity

Have one or two students present solutions to Question 1. If their approaches differ as to which part of the base is called *y,* their equations will be different, and they will get different values for *y.* Of course, they should get the same result for *k* and the same area. (*Note:* The side of length 15 splits into segments of lengths exactly 6.6 feet and 8.4 feet, and the altitude's length is exactly 11.2 feet.)

In the discussion of Question 2, elicit the idea that the area is the amount of space inside and doesn't change just because it is looked at from a different perspective. Bring out that area is not a formula but a type of measurement.

Ask, **How can you use the fact that the area doesn't change to find the value of y?** Students should realize that $\frac{1}{2} \cdot k \cdot 15$ must equal 84 and that once they know k, they can find y from one of the right triangles.

If you discuss Question 3, you might ask students to revisit it using the ideas just discussed. (*Note:* The side of length 13 splits into segments of lengths $5\frac{5}{13}$ feet [≈ 5.38 feet] and $7\frac{8}{13}$ feet [≈ 7.62 feet] and has an altitude of length $12\frac{12}{13}$ feet [≈ 12.92 feet].)

Key Question

How can you use the fact that the area doesn't change to find the value of y?

IMP Year 2, Do Bees Build It Best? Unit, Teacher's Guide

74

© 2010 Interactive Mathematics Program

The Corral Problem

Intent

In these activities, students use their earlier work with area and trigonometry to derive a method for finding the area of any regular polygon, a skill needed to tackle the unit problem.

Mathematics

The central idea driving the activities in *The Corral Problem* is the relationship between the number of sides of a regular polygon of a given perimeter and its area. This relationship is expressed as the *isoperimetric theorem,* which states that of all plane figures with a given perimeter, the circle has the greatest area.

A second important mathematical idea in *The Corral Problem* is similarity, which arises in two ways. First, students explore the relationship between the perimeter and the area of similar figures, learning that the ratio of the areas of similar polygons is the square of the ratio of their perimeters. Second, similarity is the big idea behind right-triangle trigonometry. Students use the trig functions and their inverses to find heights and angles of triangles.

The **inverse trigonometric functions** are treated as examples of the larger set of inverse functions. In addition, students explore the square-root function and its inverse.

Progression

These activities start with the corral problem, extending students' work from rectangles to regular polygons of any number of sides. Students also explore some numeric issues related to the square-root function, and they find the angles of a triangle using inverse trigonometric functions. In addition, they complete work and make presentations on *POW 2: Tessellation Pictures* and begin work on the final POW of the unit.

Don't Fence Me In

Rectangles Are Boring!

More Fencing, Bigger Corrals

More Opinions About Corrals

Simply Square Roots

Building the Best Fence

Falling Bridges

Leslie's Floral Angles

IMP Year 2, Do Bees Build It Best? Unit, Teacher's Guide

© 2010 Interactive Mathematics Program

75

Don't Fence Me In

Intent

This activity leads students to formulate and find evidence to support a speculation that of all rectangles with a given perimeter, a square has the greatest area.

Mathematics

This is the first in a series of activities that involve maximizing area for a given perimeter. Students experiment with various rectangles of a given perimeter and discover that a square has the greatest area. This discovery suggests a general principle: namely, that for a given number of sides, the most efficient polygon is a regular polygon. Proof of this principle, called the *isoperimetric theorem,* is beyond the scope of the unit.

As students are tasked with writing an expression for area in terms of width, they encounter a situation in which it is difficult to develop an algebraic expression by examining a table of values. In this case, an expression is more easily derived by mimicking the steps of their calculations.

Progression

Students work individually to compile a table of areas for various widths and to write an expression for area in terms of width. The discussion that follows centers on how the expression was developed, then examines the question of maximizing the area. Students graph the related equation to help confirm their suspicions that a square is the most efficient rectangle in this respect.

Approximate Time

25 minutes for activity (at home or in class)
15 minutes for discussion

Classroom Organization

Individuals, followed by whole-class discussion

Materials

Poster paper

Doing the Activity

This activity requires little or no introduction.

Discussing and Debriefing the Activity

Begin the discussion by having several students suggest number pairs for the table described in Question 2. Their examples will probably strongly suggest that the maximum area is achieved when length and width are equal—that is, when the rectangle is a square.

IMP Year 2, Do Bees Build It Best? Unit, Teacher's Guide

76

© 2010 Interactive Mathematics Program

You may want to bring out (perhaps by offering inappropriate *In* values) that the *In* must be a number between 0 and 150. If you discuss this, you might also review the term *domain.*

Ask for a volunteer to give a formula for the table. Finding a rule from the table itself is not easy, so ask for an explanation of how students found it. If nobody found the formula, lead the class through the area computation and suggest they mimic the numeric computation with algebra. Two likely equations are

$$A = \left(\frac{300 - 2w}{2}\right)w \ \text{ and } \ A = \left(150 - w\right)w$$

Confirming with a Graph

Students will need to change variables in order to enter the equation and will probably need to adjust the viewing window to get an appropriate view of the function. They should come up with a graph like the one here.

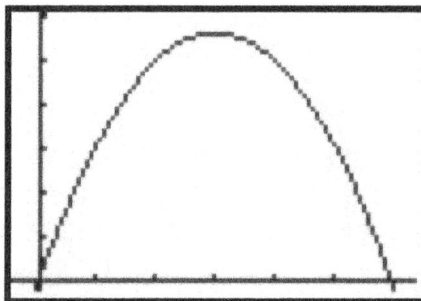

They can then use the **ZOOM** and **TRACE** features to confirm that the maximum value occurs very near *w* = 75, as next shown. The symmetry of the graph should also suggest that the maximum is halfway between 0 and 150.

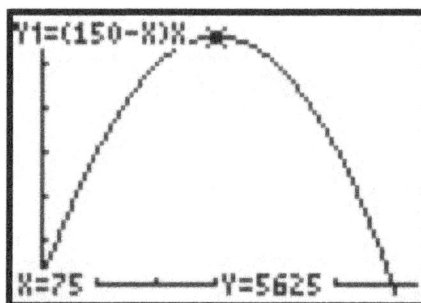

In fact, *w* = 75 does maximize the function. Post this result:

Among all rectangles with a perimeter of 300 feet, the maximum area occurs for a 75–by–75-foot square, for which the area is 5625 square feet.
Because students will be developing similar results for other polygons, you may want to enter this result on a poster as a row of an In-Out table, with "Type of polygon" as the *In* and "Maximum area" as the *Out.*

IMP Year 2, Do Bees Build It Best? Unit, Teacher's Guide

© 2010 Interactive Mathematics Program

77

In *Fireworks,* in the activity *Pens and Corrals in Vertex Form,* students will use algebra to prove that of all rectangles of a given perimeter, a square has the greatest area. In fact, a square has the maximum area of all quadrilaterals of a given perimeter, but this is harder to prove.

Connection to the Unit Problem

It may be useful to ask someone to remind the class what the central problem in this unit is—namely, finding a shape for the honeycomb cell that will give the most storage (volume) for a given amount of wax. Help students to realize that this issue of efficient use of materials is similar to the question about enclosing the most area for a given amount of fencing.

IMP Year 2, Do Bees Build It Best? Unit, Teacher's Guide

78

© 2010 Interactive Mathematics Program

Rectangles Are Boring!

Intent

Students continue the work they started in *Don't Fence Me In,* finding and then comparing the area of an equilateral triangle to that of a square with the same perimeter.

Mathematics

In *Don't Fence Me In,* students noted that of all the rectangles with a perimeter of 300 feet, a square has the most area. Now they will consider triangles. Of all triangles of a given perimeter, the equilateral triangle has the greatest area. Students are asked to compare the area of a square to that of an equilateral triangle with the same perimeter. To compute the area of the triangle, they must first find the height of the triangle. To do this, they might make use of the method derived from the Pythagorean theorem in *Leslie's Fertile Flowers* in which they decomposed the triangle into two right triangles, wrote two expressions for the altitude, and set those expressions equal.

Progression

Students will work on the activity in groups. Following the discussion, post the result for triangles with the previous results for rectangles.

Approximate Time

25 minutes

Classroom Organization

Groups, followed by whole-class discussion

Doing the Activity

In anticipation of the next activity and the subsequent activities on regular polygons, ask the class, **What triangle with a perimeter of 300 feet has the most area?** After having noticed in *Don't Fence Me In* that a square is the most efficient rectangle in this regard, students will probably suggest an equilateral triangle. Tell them that this is correct. (A complete proof is beyond the scope of this unit.)

How could we generalize the examples of squares and equilateral triangles? If necessary, review the concept of a **regular polygon** and explain that among all polygons with a given number of sides and a given perimeter, the regular polygon has the greatest area. You might also remind students of the similar conjecture they reached with regard to volume in *Building the Biggest.*

As groups work, help them one at a time or call the class together as needed. Groups are to find the area of an equilateral triangle with a perimeter of 300 feet and compare it with the area of a square with the same perimeter. The key is drawing an altitude and using the Pythagorean theorem to find its length. Try to

IMP Year 2, Do Bees Build It Best? Unit, Teacher's Guide

79

© 2010 Interactive Mathematics Program

elicit this idea from students. For example, ask, What do you need to know to find the area of a triangle?

Discussing and Debriefing the Activity

When most groups are finished, have one or two students explain their analyses. They will probably use the idea that the altitude divides the base into two equal segments of length 50 feet, as shown here.

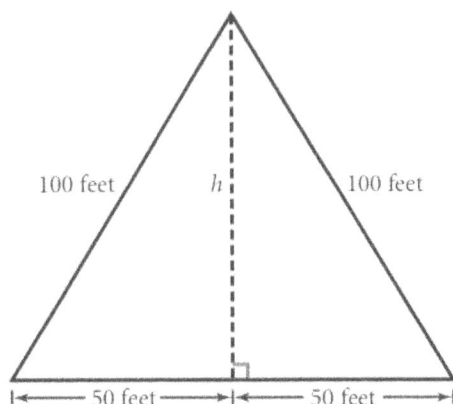

Students might notice that this diagram resembles the one from *Leslie's Fertile Flowers.* In that activity, the slanted sides of the original triangle had different lengths and the task of finding how the base split was nontrivial. The situation here is simpler because the original triangle is equilateral.

Some students may give the altitude h as $\sqrt{7500}$ while others may give a decimal approximation. In either case, ask for the area to the nearest square foot (4330 square feet).

Post this result—that the area of an equilateral triangle with perimeter 300 feet is approximately 4330 square feet—with the results for rectangles from *Don't Fence Me In,* perhaps as another row of the In-Out table. Be sure students notice that the area of the triangle is less than that of the square. They will need this result, as well as the area of the square, in *More Fencing, Bigger Corrals.*

Key Questions

What triangle with a perimeter of 300 feet has the most area?

How could we generalize the examples of squares and equilateral triangles?

What do you need to know to find the area of a triangle?

More Fencing, Bigger Corrals

Intent

In this activity, students explore the relationship between the perimeters of similar figures and their areas. They discover that the areas of similar polygons are proportional to the squares of their perimeters.

Mathematics

This activity establishes the principle that the ratio of the areas of similar polygons is the square of the ratio of their perimeters. (Some students may have encountered this idea in the supplemental activities *Fit Them Together?* and *Similar Areas* in the Year 1 unit *Shadows.* This theme is continued in *Cereal Box Sizes* and *More About Cereal Boxes.*) If two similar figures have perimeters in the ratio 2:1, for example, their areas will be in the ratio 4:1. Students use area formulas for a square and an equilateral triangle to reinforce this principle. In the process of gathering data to form this conclusion, they also practice applying the Pythagorean theorem (or using trigonometry).

Progression

Students work individually to find and compare the areas for the square and triangular corrals of *Don't Fence Me In* and *Rectangles Are Boring!* if the length of the fence is doubled or tripled. They generalize their observations and share them in a class discussion.

Approximate Time

5 minutes for introduction
25 minutes for activity (at home or in class)
10 minutes for discussion

Classroom Organization

Individuals, followed by whole-class discussion

Doing the Activity

If you are assigning this as homework, you may want to suggest that students write down for their reference the areas of the square and equilateral-triangle corrals with perimeters of 300 feet.

Discussing and Debriefing the Activity

Let students report on their results for Questions 1 and 2. They should notice that doubling the perimeter produces an area four times as big, for both square and triangular corrals. Ask the class whether they think this applies to other shapes. **Does doubling the perimeter quadruple the area for *any* shape?** They will probably agree that it does.

For Question 3, students should have found that the area for 900 feet of fencing is nine times that for 300 feet of fencing. Ask them, **How can you generalize this** result and those from Questions 1 and 2? They may say that multiplying the perimeter by some factor multiplies the area by the square of that factor. Try to elicit this principle in terms of areas of similar figures. One possible formulation is,

If two polygons are similar, then the ratio of their areas is the square of the ratio of their perimeters.

Post this result (or something equivalent).

Some students may have found the areas of the larger corrals by an intuitive understanding of this general principle. Others may have done the problems by the same analysis they used for the original problems. It is important for them to acknowledge that both approaches are correct.

Optional: Area Formulas in Terms of Perimeter
Some students may have interpreted Question 4 as calling for a general formula for the area of an equilateral triangle or square in terms of its perimeter. If so, they may be interested in trying to develop such formulas. The case of the square is easier, but they should be able to use the reasoning of the specific cases to develop a general formula for the equilateral triangle as well. Here is an outline of the development of these formulas, using P to represent perimeter and A to represent area.

For the square: Divide the perimeter by 4 to get the length of a side. Square that length to get the area. This gives $A = \left(\dfrac{P}{4}\right)^2 = \left(\dfrac{P^2}{16}\right)$.

For the equilateral triangle: Divide the perimeter by 3 to get the length of a side. Then draw an altitude to get a diagram like this.

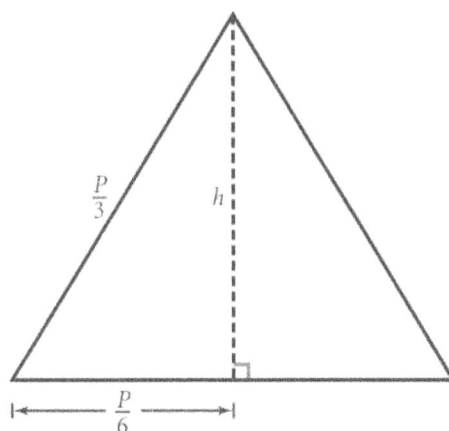

Next, use the Pythagorean theorem to get this formula for h.

$$h = \sqrt{\left(\frac{P}{3}\right)^2 - \left(\frac{P}{6}\right)^2}$$

Leaving the result in this form and applying the formula for the area of a triangle gives

$$A = \frac{1}{2} \cdot \frac{P}{3} \cdot \sqrt{\left(\frac{P}{3}\right)^2 - \left(\frac{P}{6}\right)^2}$$

For both the square and the equilateral triangle, students can verify their formulas for the case $P = 300$. For the square, they can also note that the area is a fixed multiple of P^2. Realizing this for the triangle requires simplifying the formula.

Key Questions

Does doubling the perimeter quadruple the area for *any* shape?

How can you generalize this result and those from Questions 1 and 2?

More Opinions About Corrals

Intent

Students investigate the area of a regular pentagon with a given perimeter.

Mathematics

Students have found that the area of a square with perimeter 300 feet is greater than the area of an equilateral triangle with the same perimeter. Would a regular pentagon with that perimeter enclose even more area? In this activity, students find the area of a regular pentagon with perimeter 300 feet. The challenge is to decompose the pentagon into figures they can work with and then to find the lengths they need to compute the areas of those figures. In the process, students review angle–sum relationships and trigonometric functions.

Progression

Students work in groups to find the area of a pentagonal-shaped corral. This relatively complex problem requires them to apply angle–sum relationships, trigonometry, and the formula for the area of a triangle. Expect them to need some guidance or at least an opportunity to share ideas as a class as they work. In the follow-up discussion, students compare their results with those from the earlier activities.

Approximate Time

40 minutes

Classroom Organization

Groups, interspersed with whole-class discussion

Doing the Activity

Let groups begin work on the activity. If they have no idea how to start, give them a few minutes before offering any suggestions. It may be necessary to blend group work with a whole-class discussion of the key steps.

To help groups get started, you might ask, Can you break the pentagon into pieces that are easier to work with? They might recognize that they can split the pentagon into five triangles as shown next.

IMP Year 2, Do Bees Build It Best? Unit, Teacher's Guide

84

© 2010 Interactive Mathematics Program

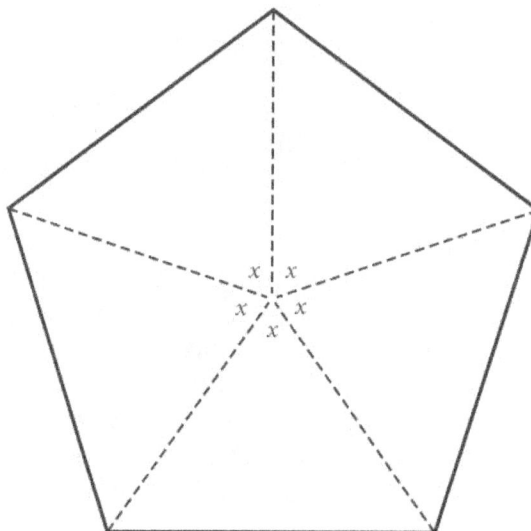

Use the term *central angle* for the equal angles (labeled *x*) at the center, but explain that this term is usually used for angles formed by radii of a circle. Students should be able to figure out the size of the central angles, as each is one-fifth of 360°. That is, *x* is 72°. We will refer to each of the triangles in this diagram as a *central triangle,* though there is no standard term for such triangles.

After students have divided their pentagons into five central triangles, ask them to work out more details about the triangles. They should realize that each triangle looks like this.

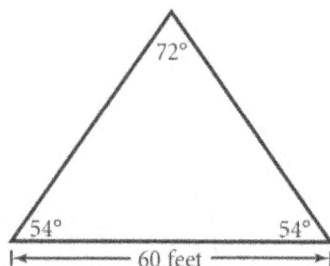

The base angles are each half of the interior angles of the pentagon. As the goal for students is to find the area of the pentagon, this suggests they need to find the length of an altitude.

You may want to bring the class together to summarize the discoveries made up to this point, even drawing in the altitude. You might check that everyone understands the source of the information in the following diagram.

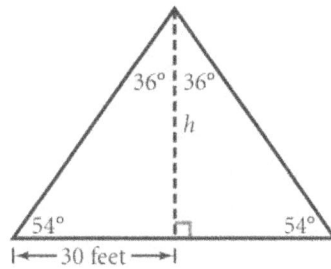

It's harder to find this altitude than it was for the equilateral triangle because students know only the length of the base and not the lengths of the other two (equal) sides. You might refer students to their earlier review of trigonometry to help them figure out how they can find the length of the altitude.

In this case, the most useful ratio for finding the altitude is the tangent function. We have

$$\tan 54° = \frac{h}{30}$$

so $h = 30 \tan 54°$. Some students may work with the 36° angle instead. If so, they will likely get

$$\tan 36° = \frac{30}{h}$$

From here, you can let students return to their groups to finish the activity. They will probably use their calculators to evaluate tan 54° and then use the expression $\frac{1}{2}bh$ to compute the triangle's area. They do need to remember that the pentagon consists of five of these triangles.

Students may try other ways of dividing the pentagon, including those here. Such divisions can be used but are more difficult to work with than the central-angle approach because they each involve more than one shape. They are also less desirable for the mathematical development of this unit because they do not generalize to other regular polygons the way the method of central triangles does.

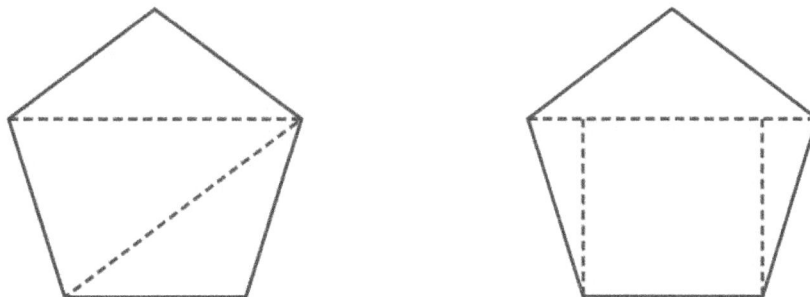

Discussing and Debriefing the Activity

It's probably a good idea to have at least one presentation summarizing the ideas. Be sure the discussion includes the term *central angle.*

When the class has reached a consensus, post this result—that the area of a regular pentagon with perimeter 300 feet is approximately 6194 square feet—along with those from the earlier activities, and have students compare the areas of the various regular polygons that have the same perimeter.

Key Question

Can you break the pentagon into pieces that are easier to work with?

Simply Square Roots

Intent

In using the Pythagorean theorem, students have been working with square roots. They now examine conjectures about the behavior of square roots and explore simplifying square-root expressions.

Mathematics

Students will explore general properties of the square-root function by examining examples of sums, products, and quotients of square roots. The square root of a product (or quotient) is equal to the product (or quotient) of the square roots, but the square root of a sum is not equal to the sum of the square roots. Thinking that the square root of a sum *is* equal to the sum of the square roots is a common misconception. These properties of the square-root function allow exact representations of irrational numbers, written as radical expressions, to be expressed in simpler form.

Progression

Students work individually to test specific numbers and to decide whether the square root of a sum, product, or quotient equals the sum, product, or quotient, respectively, of the square roots. Students apply their conclusions to simplify square-root expressions. The properties are reviewed in a class discussion, which also covers the topic of decimal approximations of square-root expressions.

Approximate Time

25 minutes for activity (at home or in class)
15 minutes for discussion

Classroom Organization

Individuals, then groups, followed by whole-class discussion

Doing the Activity

Remind students to read the instructions in the introductory paragraphs carefully before answering the questions.

Discussing and Debriefing the Activity

You might ask various groups to prepare presentations for Questions 1 through 3 and the examples in Question 4.

For Question 1, numeric examples will show that the square root of a sum is not equal to the sum of the corresponding square roots. There isn't much need to look for an explanation of this important result.

For Questions 2 and 3, however, students should determine that the square root of a product or quotient is equal to the corresponding product or quotient of square roots. Ask them to explain this fact. For example, they should understand that by definition, if they square $\sqrt{a \cdot b}$, they get $a \cdot b$. They need to recognize that if they square $\sqrt{a} \cdot \sqrt{b}$, they also get $a \cdot b$. This may take some guidance, such as with a question like, **What does it mean to square something?** This should help them consider $\left(\sqrt{a} \cdot \sqrt{b}\right)^2$ as $\sqrt{a} \cdot \sqrt{b} \cdot \sqrt{a} \cdot \sqrt{b}$; you can then help them realize how to rearrange this to show that it is equal to $a \cdot b$.

Questions 4a and 4b reflect Question 2, but you may need to guide students in rewriting the expressions "more simply" as $5 \cdot \sqrt{3}$ and $7 \cdot \sqrt{5}$. Tell students that it is convention, though not universal, that $5 \cdot \sqrt{3}$ is considered simpler than $\sqrt{75}$. Explain that the multiplication sign is usually omitted in such expressions, so $5\sqrt{3}$ and $7\sqrt{5}$ are shorthand for $5 \cdot \sqrt{3}$ and $7 \cdot \sqrt{5}$.

For Question 4c, you may need to prompt students to factor 18 in a way that will allow a similar simplification.

The upcoming activity *Falling Bridges* will point out issues with using approximations instead of exact values.

Key Question

What does it mean to square something?

IMP Year 2, Do Bees Build It Best? Unit, Teacher's Guide

© 2010 Interactive Mathematics Program

89

Building the Best Fence

Intent

This activity completes the sequence of corral problems as students develop a formula for the area of a regular n-gon with perimeter 300 feet.

Mathematics

A regular n-sided polygon, or n-gon, can be decomposed into n congruent "central triangles." For each of these triangles, students can determine the vertex angle, the base angles, the length of the base, the height, and ultimately the area. The area of the entire polygon would be n of these areas.

After developing a formula for the area of a regular n-gon with perimeter 300 feet, students recognize that the more sides a regular polygon has, the greater its area for a fixed perimeter. This leads to the conclusion that for a fixed perimeter, the circle is the shape with the most area.

Progression

Groups begin with practice in finding the area of a regular polygon, using the procedure they developed for a pentagon in *More Opinions About Corrals,* and then they generalize to the case of a regular n-gon with a perimeter of 300 feet. In the follow-up discussion, it is useful to have the class further generalize this finding to the case with an arbitrary perimeter P.

In addition, students can explore what happens as the number of sides increases. One approach to this is graphing the function and examining the graph to note what it reveals about the best shape to use for the corral. For a given perimeter, the area of a regular polygon increases as the number of sides increases, which leads to the conclusion that a circle has the greatest possible area.

Approximate Time

75 minutes

Classroom Organization

Groups, followed by whole-class discussion

Doing the Activity

In this activity, students finally consider the general problem of the area of a regular polygon with a given perimeter. Most groups will have to compute the areas of at least one or two other regular polygons, in addition to the case of the regular pentagon in *More Opinions About Corrals,* before they are able to generalize.

Groups may struggle with trying to sketch a regular polygon with a large number of sides. Remind them that as long as they can visualize the polygon, they need to draw only one of the triangles that the polygon will be divided into.

IMP Year 2, Do Bees Build It Best? Unit, Teacher's Guide

© 2010 Interactive Mathematics Program

90

If many groups are getting stuck, you might let groups that are having some success report on their specific examples. Or you might review the pentagon example with the entire class.

Discussing and Debriefing the Activity

Let a student present one group's general analysis, following up with additional presentations if other groups approached the problem differently. These steps show one way to proceed to the general formula for the area of a regular n-gon with perimeter 300 feet.

1. Find the length, l, of one side of the polygon.

$$l = \frac{300}{n}$$

2. Think of the regular n-gon as divided into n triangles, each with a vertex at the center of the polygon. Then find the size, x, of a central angle.

$$x = \frac{360°}{n}$$

3. Use the fact that the sum of the angles of a triangle is 180° to find the size of a base angle, y, of one of the individual triangles.

$$y = \frac{1}{2}\left(180° - \frac{360°}{n}\right)$$

4. If desired, simplify the expression for y.

$$y = 90° - \frac{180°}{n}$$

5. Draw the altitude for one of the individual triangles, creating two right triangles. Each of these right triangles has a side of $\frac{l}{2}$ and a base angle of $90° - \frac{180°}{n}$.

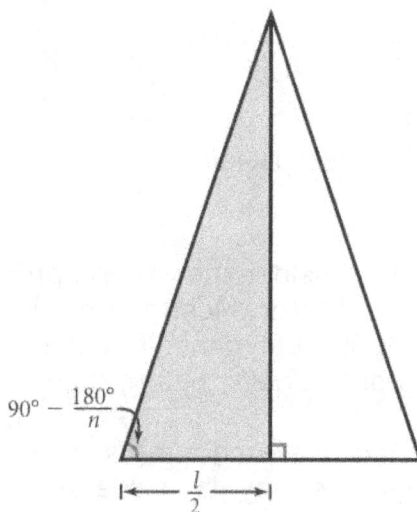

6. Using trigonometry, write an equation involving the altitude h for one of the individual triangles.

$$\tan\left(90° - \frac{180°}{n}\right) = \frac{h}{l/2}$$

7. Solve for h, and use the fact that $\frac{l}{2} = \frac{150}{n}$ to get

8.

$$h = \frac{150}{n}\tan\left(90° - \frac{180°}{n}\right)$$

9. Use the area formula for triangles to calculate that the area of one of the triangles is $\frac{1}{2}lh$.

10. Use the expressions for l and h and the fact that there are n triangles to get a formula for the area of the regular n-gon.

$$\text{Area} = n \cdot \frac{1}{2} \cdot \frac{300}{n} \cdot \frac{150}{n} \cdot \tan\left(90° - \frac{180°}{n}\right)$$

11. If desired, simplify this formula.

$$\text{Area} = \left(\frac{22500}{n}\right) \cdot \tan\left(90° - \frac{180°}{n}\right)$$

Optional: Generalizing to an Arbitrary Perimeter

You may want to have the class generalize the formula they found to consider an arbitrary value P for the original perimeter. Working from the formula in step 9 above, help them realize that P can replace 300 and $\frac{P}{2}$ can replace 150. Simplifying gives the formula

$$\text{Area} = \left(\frac{P^2}{4n}\right) \cdot \tan\left(90° - \frac{180°}{n}\right)$$

Summing Up Corrals

Now that students have a general formula for the area of a regular n-gon with perimeter 300 feet, ask them to investigate what happens as the number of sides changes. There are several approaches they might use. Use the ideas presented here to help groups as they work and make their presentations. If no one suggests it, raise the possibility of using a graph to explain what's happening.

Trying specific values: Students might begin with the obvious approach—simply substituting various values for n. Although this is a legitimate start, encourage them to look for an approach that gives an overall picture of what's happening.

Graphing: Students might graph the function

$$A = n \cdot \frac{1}{2} \cdot \frac{300}{n} \cdot \frac{150}{n} \cdot \tan\left(90° - \frac{180°}{n}\right)$$

on their calculators, with appropriate changes in variables as required. Getting an appropriate viewing window for this function will involve some consideration. They should realize that *n* should be at least 3 for this to make sense. For a pentagon, the area was almost 6200 square feet, so they might want to set the window to include values for *A* up to, say, 10,000. The function is graphed here.

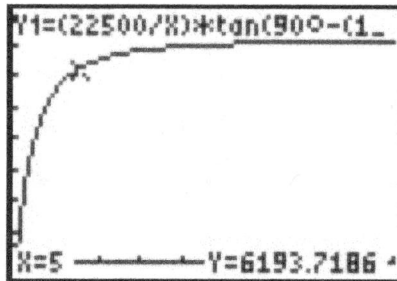

Some students will notice that increasing the number of sides makes the polygon more and more like a circle. Give them the formulas for circumference and area of a circle if they ask for them so they can verify their intuition about what shape has the greatest area.

If students question whether it makes sense to graph this function, as the number of sides must be a whole number and the graph allows any value for *n*, let them discuss the issue.

Using a table: Students can use the calculator's **TABLE** feature to get the areas for special values of *n*. They may find, though, that using a table makes it difficult to jump to much larger values.

So What's the Best Shape?
Whatever approaches groups use, they will presumably all figure out that the greater the number of sides, the greater the area. Be sure not to leave the discussion without asking, **What is the best shape for rancher Gonzales's corral?** Some students will probably conclude that it is a circle, realizing that a

IMP Year 2, Do Bees Build It Best? Unit, *Teacher's Guide*

© 2010 Interactive Mathematics Program

93

circle is "the ultimate many-sided polygon." If no one suggests a circle, ask students to draw a 100-sided polygon.

The IMP curriculum does not work with circles formally until Year 3, but you may want to calculate the area of a circle with circumference 300 feet. You can briefly describe the formulas for the computation, which gives an area of approximately 7162 square feet.

Once the class recognizes that the shape with the most area is a circle, you might ask whether the rancher should build her corral in that shape. They might point out that many fence materials don't lend themselves to circles and that in practical terms, there isn't much difference between a regular polygon with, say, 20 sides, and a circle.

Key Questions

What is the best shape for rancher Gonzales's corral?

Supplemental Activity

Toasts of the Round Table (extension) involves geometric considerations similar to those in the corral problems.

Falling Bridges

Intent

While solving equations derived from the Pythagorean theorem, students have been dealing with expressions involving square roots. In *Simply Square Roots,* they explored some properties of the square-root function. Now they examine the consequences of rounding errors in computations involving square roots.

Mathematics

This activity provides a vivid illustration of the importance of retaining the full accuracy of an answer in radical form by demonstrating that rounding an intermediate answer can have an enormous effect on the final result. This activity will also help make irrational numbers seem more relevant and less abstract.

Progression

Students work on the activity individually and compare results in their groups and in a brief class discussion.

Approximate Time

15 minutes for activity (at home or in class)
10 minutes for discussion

Classroom Organization

Individuals, then groups, followed by whole-class discussion

Doing the Activity

You may want to remind students that $70\sqrt{2}$ means 70 times $\sqrt{2}$.

Discussing and Debriefing the Activity

Let students compare results in their groups while you get a feeling for how well they did with this activity. If the results seem good, you might have one or two students make presentations.

If not many students understood how the figures got so far off, ask them to calculate the safe load, first using $\sqrt{2} \approx 1.4$ [this gives 1000(99 – 70 • 1.4), or 1000] and then using $\sqrt{2} \approx 1.41$ [this gives 1000(99 – 70 • 1.41), or 300]. Students may be surprised at the effect of using one additional decimal place. Let them investigate the consequences of using closer and closer approximations for $\sqrt{2}$.

Here are results of using successively closer approximations for $\sqrt{2}$.

1000(99 – 70 • 1.4) = 1000   **1000(99 – 70 • 1.414213) = 5.09**
1000(99 – 70 • 1.41) = 300   **1000(99 – 70 • 1.4142136) = 5.048**
1000(99 – 70 • 1.414) = 20   **1000(99 – 70 • 1.41421356) = 5.0508**
1000(99 – 70 • 1.4142) = 6   **1000(99 – 70 • 1.414213562) =**
1000(99 – 70 • 1.41421) = 5.3   **5.05066**

IMP Year 2, Do Bees Build It Best? Unit, Teacher's Guide

© 2010 Interactive Mathematics Program

95

Thus the correct value based on Thorough Ted's work is roughly 5.05 tons, not the 1000 tons Speedy Sam calculated.

Have students use the distributive property to expand the expression $1,000\left(99 - 70\sqrt{2}\right)$ to get the expression $99,000 - 70,000\sqrt{2}$. Then ask, How does this expanded expression explain why changing the value used for $\sqrt{2}$ from 1.414 to 1.4142, which is a change of only 0.0002, makes such a big difference in the final answer?

Key Question

How does this expanded expression explain why changing the value used for $\sqrt{2}$ from 1.414 to 1.4142, which is a change of only 0.0002, makes such a big difference in the final answer?

Supplemental Activity

What's the Answer Worth (reinforcement) is a good follow-up to this activity. When students measure side lengths, their answers are always approximations. In this activity, they look at the impact of rounding errors on their final answers.

IMP Year 2, Do Bees Build It Best? Unit, Teacher's Guide

© 2010 Interactive Mathematics Program

96

Leslie's Floral Angles

Intent

This new look at the situation introduced in *Leslie's Fertile Flowers* gives students a chance to apply the basic ideas of trigonometry again and provides an opportunity for introducing the inverse trigonometric functions.

Mathematics

When we know the side lengths of a right triangle, we can use trigonometry to find the acute angles of that triangle. In this activity, students repeatedly apply the basic trigonometric functions to find these acute angles. This repetition provides motivation for the subsequent introduction of the **inverse trigonometric functions.**

Progression

Working individually, most students will accomplish this task using a guess-and-check procedure, repeatedly applying the sine, cosine, and tangent functions. The subsequent class discussion will build on this experience as motivation for the introduction of the inverse trigonometric functions, including discussion of notation and domain for them.

Approximate Time

15 minutes for activity (at home or in class)
25 minutes for discussion

Classroom Organization

Individuals, followed by whole-class discussion

Doing the Activity

This activity will require little or no introduction.

Discussing and Debriefing the Activity

It will help the discussion if you have the class label the vertices of the various triangles, as shown here.

IMP Year 2, Do Bees Build It Best? Unit, Teacher's Guide

97

B

13   12   15

A   5   D   9   C

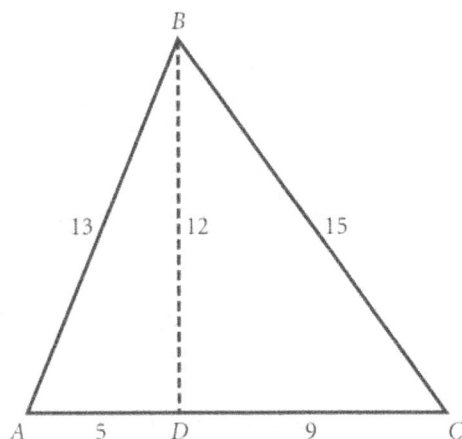

Have several volunteers present their ideas on how to find the angles. They will probably start with trigonometric equations. For example, students could use any of these equations to determine $\angle A$.

$$\tan A = \frac{12}{5} \qquad\qquad \sin A = \frac{12}{13} \qquad\qquad \cos A = \frac{5}{13}$$

Students probably found the value of $\angle A$ from one of these equations by guessing and checking, plugging in values for A until they got reasonably close to what they wanted.

To find $\angle ABC$, students may have found both $\angle ABD$ and $\angle DBC$ from the right triangles. Or they may have found $\angle A$ and $\angle C$ and subtracted their sum from 180°. You might suggest using one of these methods to check the other.

To the nearest degree, $\angle A = 67°$, $\angle B = 59°$, and $\angle C = 53°$. Due to rounding, these values don't add to 180°.

The Inverse Trigonometric Functions

This discussion begins with the inverse tangent function, because for acute angles, the tangent function has a range of all positive numbers. (The term *range* was introduced in the Year 1 unit *Patterns,* but students may need a reminder of its meaning.) If students' work on the activity suggests using sine or cosine first, make appropriate adjustments. Additionally, this discussion of inverse trigonometric functions considers only acute angles.

Have students find the tangent of several angles between 0° and 90° to review the fact that the range of the tangent function is the set of all positive numbers. You might ask them to explain this fact. For instance, they might talk about "tall, skinny right triangles" and "short, wide right triangles" to illustrate extreme values for the tangent function.

Tell students that for any positive number t, the angle whose tangent is t is called the *inverse tangent of t* and is written $\tan^{-1} t$.

Illustrate this notation by asking what students think $\tan^{-1} 0.84$ means. Have them find the $\tan^{-1}$ key on their graphing or scientific calculators and use it to confirm that $\tan^{-1} 0.84 \approx 40°$. For further practice, let them find $\angle A$ and $\angle C$ from *Leslie's Floral Angles* using this key.

Caution: If students use the fraction $\dfrac{12}{5}$ for the ratio of opposite side to adjacent side, they may need to insert parentheses to get the inverse tangent. For example, to find $\angle A$, they may need to enter $\tan^{-1} (12/5)$. If they enter $\tan^{-1} 12/5$ (without parentheses), their calculators may find $\tan^{-1} 12$ and divide the result by 5.

Tell the class to think of $\tan^{-1}$ as a single symbol, without worrying why it is written as though it had an exponent of -1. The -1 is not functioning as an exponent here.

Inverse Tangent and Right Triangles
Bring out that the definition of the tangent function as a ratio of sides is based on having a right triangle. The tangent of an angle in a nonright triangle is generally not equal to a ratio of sides of that triangle.

The Concept of an Inverse Function
Before going on to the other inverse trigonometric functions, spend a few minutes discussing the general idea of an inverse function, perhaps by comparing it to the concept of square root.

Ask students, **What do you call the number whose square is 7?** They should know that it is called the square root of 7 and is written $\sqrt{7}$. **What happens if you square the square root of 7?** If necessary, have them enter $\sqrt{7}$ on their calculators and square the result to note that $\left(\sqrt{7}\right)^2 = 7$. Also ask what happens if they start with some number, square it, and then take the square root. Have them test their hypotheses. They might realize, for example, that $\sqrt{13^2} = 13$.

Explain that in a sense, the inverse tangent function is similar to the square-root function. Just as $\sqrt{7}$ means "the number whose square is 7," so $\tan^{-1} 2$ means "the angle whose tangent is 2."

What do you think you get if you take the tangent of this number: $\tan^{-1} 2$? Have students do this, finding $\tan^{-1} 2$ and taking the tangent of the result. They should realize that $\tan (\tan^{-1} 2) = 2$.

Similarly, have them start with an acute angle, take its tangent, and then find $\tan^{-1}$ of that number. They should observe that they get their original angle (except perhaps for rounding error).

Emphasize that in using the $\tan^{-1}$ key, they give the calculator a number (which represents a ratio of the legs of a right triangle) and the calculator gives them a result that represents an angle. In contrast, when they use the tan key, they give the calculator an angle and get a number.

Inverse Sine and Cosine

When students have a reasonably clear idea of what the inverse tangent function means and how the tan^{-1} key works, you can likely get them to develop the corresponding ideas for the sine and cosine. You might use the diagram from *Leslie's Floral Angles* to point out the relationship $\sin A = \frac{12}{13}$ and ask, **How could you get the value of angle *A* from the equation** $\sin A = \frac{12}{13}$? They will probably be able to give you the appropriate terminology and notation.

Use a few sample angles to review the fact that the range of the sine function is the set of numbers between 0 and 1. **What does this fact say about the domain of the inverse sine function?** Have students confirm that numbers greater than 1 don't work with the inverse sine function by having them attempt to use the sin^{-1} key in an expression such as sin^{-1} 2. Their calculators should give them an error message.

If students raise the question of whether the values 0 and 1 themselves are in the range of the sine function, you will be faced with a bit of a pedagogical dilemma. Although sin 0° = 0 and sin 90° = 1, these facts do not quite fit the right-triangle definition students know. You can tell them that the trigonometric functions can be extended to include angles that are not strictly between 0° and 90° and have them confirm the particular cases of sin 0° and sin 90° on their calculators. (They will learn about the broader definitions of the trigonometric functions in the Year 3 unit *High Dive.*)

You can use a similar approach, perhaps somewhat abbreviated, to introduce the inverse cosine function.

Key Questions

Why can the output from the tangent function be any positive number?

What do you call the number whose square is 7?

What happens if you square the square root of 7?

What do you think you get if you take the tangent of this number: tan^{-1} 2?

How could you get the value of angle *A* from the equation $\sin A = \frac{12}{13}$?

So what is the domain of the inverse sine function?

IMP Year 2, Do Bees Build It Best? Unit, *Teacher's Guide*

© 2010 Interactive Mathematics Program

100

From Two Dimensions to Three

Intent

Up to this point in the unit, students have been working in two dimensions. But the unit problem—determining whether the hexagonal honeycomb is the most efficient shape for storing honey—concerns three dimensions. These activities extend the ideas developed so far to three-dimensional figures.

Mathematics

Polyhedra are three-dimensional figures whose faces are polygons. **Prisms** are a particular type of polyhedron. They have parallel polygonal bases and faces that connect corresponding edges.

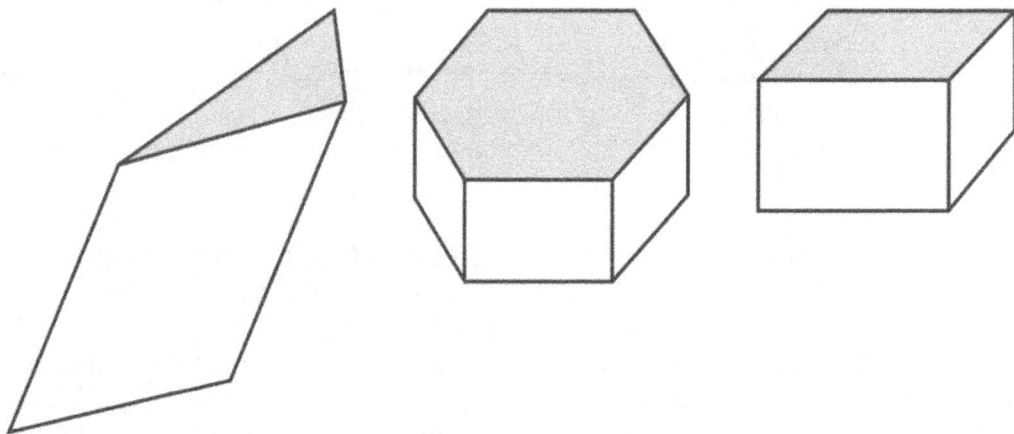

The lateral faces of prisms are parallelograms. Right prisms have lateral sides perpendicular to the bases. The volume of a prism is the area of a base multiplied by the prism's height. The lateral surface area of a prism is the perimeter of a base multiplied by the prism's height.

In *From Two Dimensions to Three,* students will develop ways to find the volumes and surface areas of prisms, skills that will be useful for solving the unit problem, which involves right hexagonal prisms. In addition, they will extend the Pythagorean theorem to three dimensions by finding a formula for the length of the interior diagonal of a rectangular prism, as well as examine surface areas and volumes for similar polyhedra.

Progression

The activities begin with explorations of surface area and volume for rectangular solids. Prisms are then defined, and students use patterns to find formulas for their volumes and surface areas. In addition, students complete the final POW of the unit.

POW 3: Possible Patches

Flat Cubes

IMP Year 2, Do Bees Build It Best? Unit, Teacher's Guide

© 2010 Interactive Mathematics Program

101

POW 3: Possible Patches

Intent

This POW gives students practice in communicating technical information in a clear, organized fashion.

Mathematics

This POW is peripherally concerned with the topic of area. Additionally, as with all POWs, it will help students become accustomed to approaching difficult problems from a number of directions and give them an opportunity to practice organizing and communicating complex ideas.

Progression

Students examine how to cut out as many copies of a given shape as possible from a given piece of material.

Approximate Time

5 minutes for introduction
1–3 hours for activity (at home)
15 minutes for presentations

Classroom Organization

Individuals, followed by whole-class presentations

Doing the Activity

Have students read the introduction to the POW. Emphasize that their first task is to find an explicit way to cut the piece of satin to create the maximum number of patches they think possible, rather than merely stating a number. Their write-ups should include diagrams of their methods.

Some students may think they can find the maximum number of 3-by-5-inch patches in a 17-by-22-inch piece of satin by simply dividing the area of the satin (374 square inches) by the area of the patch (15 square inches). Although this POW asks students to describe a *specific* method for cutting the satin, it would be helpful to eliminate this possible confusion early on.

You might clarify this issue by asking, **How many patches could Keisha cut from a 2-by-30-inch piece of satin?** Someone will likely point out that although the area of the satin is 60 square inches, no 3-by-5-inch patches could be cut from it. Point out that this is one of the reasons students are asked to draw a diagram showing how to cut the satin.

On the day before the POW is due, choose three students to make presentations and give them transparencies and pens to take home to use in their preparations. Ask them not only to prepare diagrams of their solutions, but also to prepare a

IMP Year 2, Do Bees Build It Best? Unit, Teacher's Guide

103

discussion of how they approached the problem. You might suggest they each present a different part of the POW and decide among themselves how to do this.

Discussing and Debriefing the Activity

Ask the three students to make their presentations.

Students should have found that Keisha couldn't get more than 24 patches for Question 1, so the focus should be on exactly how they determined the maximum. **How do you know that you have found the maximum number of patches possible?** Try to encourage more details about their solution processes than "I just kept moving the patches around until I got it." Students in the audience may have valuable contributions to make to this discussion.

For Questions 2 and 3, focus on how students know they have found the maximum number of patches possible. They may not be able to offer airtight proofs, but they should be able to give reasonably convincing arguments.

Key Questions

How many patches could Keisha cut from a 2-by-30-inch piece of satin?

How do you know that you have found the maximum number of patches possible?

IMP Year 2, Do Bees Build It Best? Unit, Teacher's Guide

104

© 2010 Interactive Mathematics Program

Flat Cubes

Intent

The concept of surface area is introduced and then reinforced through the construction of nets for a cube.

Mathematics

The unit problem is framed in terms of maximizing storage capacity—and capacity, or **volume,** is a three-dimensional concept. Students now turn to an examination of volume and to the extension of area to three-dimensional objects through the concept of **surface area**. This activity sets up further consideration of surface area by asking students to visualize how a two-dimensional pattern called a **net** is transformed into a three-dimensional object.

Progression

A class discussion reviews the status of the unit problem and introduces the idea of surface area. Students work on the activity individually and share findings in a class discussion.

Approximate Time

10 minutes for introduction
25 minutes for activity (at home or in class)
10 minutes for discussion

Classroom Organization

Individuals, followed by whole-class discussion

Materials

One-Inch Graph Paper blackline master (2 sheets per student)

Doing the Activity

This activity addresses the concept of surface area, a new aspect of the unit problem. In anticipation of this, take a few minutes to reorient students to the honeycomb problem. You might begin by asking for volunteers to address this question: Once again, what is the central problem of the unit? How is it related to the corral problem? Students should recognize, in some general way, that both topics relate to making the best use of materials.

- In the corral problem, rancher Gonzales wants the most possible space for her horses for a given amount of fencing.
- In the honeycomb problem, the bees "want" the most possible space for storing honey for a given amount of wax.

Help students to articulate that in the corral problem, the fixed item is perimeter and the thing to be maximized is area. Ask, What are you trying to maximize in the honeycomb problem? (volume) What is being kept fixed? (amount of wax). Explain that students should assume the walls of the honeycomb are thin and that they can think of the "fixed amount of wax" so that the wax forming the walls

IMP Year 2, Do Bees Build It Best? Unit, Teacher's Guide

105

can be treated as if it were like the sheets of construction paper students used in *Building the Biggest.* Thus, the quantity of wax is essentially measured by examining the surface area of its walls. Introduce the term *surface area* now, although it is not necessary to formally define it at this time. Also explain that although honeycomb cells are open at one end and closed at the other, students will investigate a simplified problem—that is, they will ignore the wax used for the "ends" of the cells.

Discussing and Debriefing the Activity

Begin the discussion by randomly choosing some examples from students' work and asking the class or groups to decide whether each example is a net for a cube. In addition, you might make transparencies of the outlines here and ask groups to investigate whether they are nets for a cube.

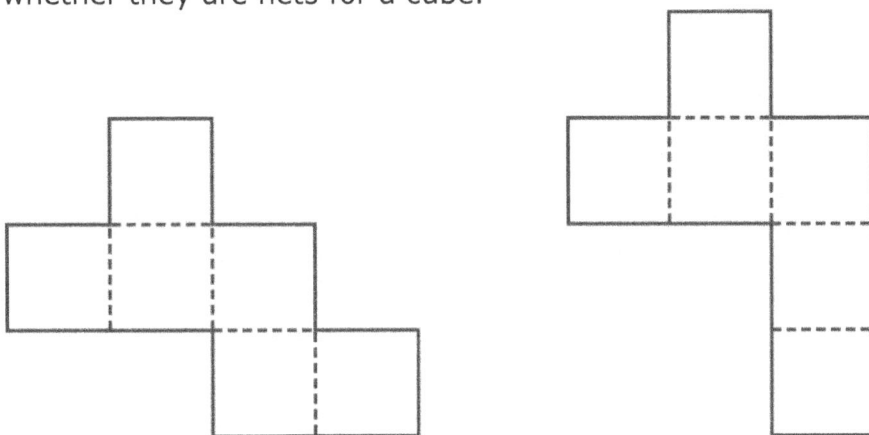

Ask students to try to decide whether each example is a net for a cube *before* they cut and fold to test their ideas. (It is not always easy to imagine whether such diagrams will fold into a cube; three-dimensional visualization is a complex skill. The first outline above is a net for a cube; the second is not.)

Have the class compile as complete a set of possible nets for a cube as they can find. Keep asking for new possibilities until all their ideas have been shared. You can leave open the question of whether all the possibilities have been found.

This process may lead to a discussion as to what constitutes "different" nets. For example, students may disagree as to whether the mirror image of a given net is different or not. Use your judgment as to how far to go with this discussion, perhaps mentioning that the definition of what is "different" is sometimes a matter of convention, not pure logic.

Key Questions

Once again, what is the central problem of the unit? How is it related to the corral problem?

What are you trying to maximize in the honeycomb problem? What is being kept fixed?

Flat Boxes

Intent

Students extend their work in *Flat Cubes* by making nets for and finding the surface areas of rectangular prisms.

Mathematics

Students generalize the ideas from *Flat Cubes,* extending the work with cubes to more general rectangular solids. The surface area of a rectangular solid can be thought of as the area of a net that can be used to create it. Thus the net is a physical model that helps students transition from two dimensions to three.

Progression

The activity is introduced with a brief discussion of the concept of surface area and a review of the terms **face, edge,** and **rectangular solid.** Then, in pairs, students make nets for rectangular solids that are not cubes and use them to find surface areas. Pairs share their findings in a class discussion. Students save their nets for use in a future activity.

Approximate Time

40 minutes

Classroom Organization

Pairs, followed by whole-class discussion

Materials

One-Centimeter Graph Paper blackline master (2 sheets per pair of students)

Doing the Activity

Begin by mentioning that many shapes besides cubes can be built from nets and that today the class will make nets for other figures.

Ask students to find the areas of their nets for the cube from *Flat Cubes.* They should find that all the nets have an area of 6 square inches. Ask, **Why do all the nets have areas of 6 square inches?** They should say something like, "The cube has six faces, and each face is a 1-inch square." If students don't use the term **face,** introduce it here. (The term has been used in several previous units, though informally, as in the phrase "face of a die.")

Review the term **surface area** (introduced informally with *Flat Cubes*) to describe this aspect of measuring the cube. For example, you might state that the surface area of the cube is 6 square inches. Let students try to articulate the meaning of the term; it is not an easy concept to express in words. One way to think of surface area is "the area a solid would have if its 'outside' were opened up and laid out flat." Students might also consider the case of a solid with polygonal faces. Bring out the idea that surface area is the sum of the areas of all the faces.

IMP Year 2, Do Bees Build It Best? Unit, Teacher's Guide

107

Point out that a net offers an easy way to look at surface area, because it turns a three-dimensional surface into a two-dimensional figure.

Also introduce the term **edge** and ask, How many edges does a cube have? You might point out that each face has four edges and that the cube has six faces, and then ask why the cube does not therefore have 6 • 4, or 24, edges.

Tell students that after cubes, the next simplest category of solids is boxes, more formally called *rectangular solids.* (We will use the two terms interchangeably.) Ask students to explain what they mean by a box. Their description should include these features.

- A box has six faces.
- The faces of a box are all rectangles.
- Adjacent faces of a box are perpendicular.
- Opposite faces of a box are parallel.

Make sure students realize that a cube is a particular type of box, just as a square is a particular type of rectangle.

Have students work on the activity in pairs.

Discussing and Debriefing the Activity

Let several pairs present their nets for each of the boxes and describe how they found the surface areas. Emphasize that as with the cube, there may be several different nets for a given box, but they all have the same area and are made up of six rectangles corresponding to the six faces of the rectangular solid.

This is a good opportunity to review the concept of units for area and the idea that in terms of square units, area answers the question, "How many squares fit inside?"

Have students save their boxes (or flatten the boxes and save them as nets) for use in *The Ins and Outs of Boxes.*

Key Questions

Why do all the nets of our cubes have areas of 6 square inches?

How many edges does a cube have?

Not a Sound

Intent

Students measure and find the surface area of a large rectangular solid. They also consider again how surface area is related to the unit problem.

Mathematics

The concept of surface area is reinforced through practice in finding the surface area of a room in a house.

Progression

Each student works alone to measure a room at home and to calculate the quantity of soundproofing material that would be needed to cover the walls, floor, and ceiling. The subsequent class discussion will center on the measurements that were necessary, as well as the calculation of surface area.

Approximate Time

20 minutes for activity (at home or in class)
5 minutes for discussion

Classroom Organization

Individuals, followed by whole-class discussion

Doing the Activity

This activity requires little or no introduction.

Discussing and Debriefing the Activity

Begin the discussion by having volunteers share how they measured the dimensions of their rooms, including the measurement units they used, the tools they used, and the number of measurements they made to be able to calculate the amount of soundproofing.

If none of the presenters uses the fact that the "faces" of the "boxes" came in pairs (floor and ceiling, left and right, front and back), ask whether anyone found a shortcut for computing the amount of soundproofing.

Key Questions

Was one unit more convenient than another? Why?

Did you use a ruler? A tape measure? Estimation?

Did you measure all the edges of your "box"?

Did you find any shortcuts?

IMP Year 2, Do Bees Build It Best? Unit, Teacher's Guide

109

A Voluminous Task

Intent
This activity begins a systematic treatment of the concept of volume.

Mathematics
The concept of **volume** is introduced through a discussion of units for measuring volume. In this first activity with the concept, students build solids composed of unit cubes and use these physical models to determine the volumes and surface areas of the solid figures. These concrete examples help students focus on the idea of volume and surface area as "how many units" rather than in terms of formulas.

Progression
The activity is introduced through discussion of the meaning of volume and of units for measuring volume. Students then work in groups to build and then find the volumes and surface areas of figures made from cubes. No follow-up discussion will be needed.

Approximate Time
45 minutes

Classroom Organization
Groups

Materials
Linking cubes or other cubes, if linking cubes are unavailable (at least 34 per group)

Doing the Activity
In both this activity and *Shedding Light on Prisms,* linking cubes will be helpful for building stable figures. If these are not available, any uniform set of cubes will do.

Remind students that their recent work has focused on surface area and that today they will look at volume. Ask for volunteers to try to describe what volume is. What does *volume* mean? How is it different from surface area? As with surface area, this is a difficult concept to put into precise words, so general descriptions are expected here.

You might then ask students to think back to their work on area early in the unit and to do some focused free-writing on the topic, "How do you measure volume?" Let students share their ideas. They should come up with these general observations.

- Volume needs a unit.
- The unit for volume should be some sort of solid object.

IMP Year 2, Do Bees Build It Best? Unit, Teacher's Guide

110

© 2010 Interactive Mathematics Program

- One measures the volume of an object by counting how many copies of the unit are needed to "fill up" the object without gaps.
- A cube would make an excellent unit.

Let students suggest specific units for volume, such as cubic inch, cubic centimeter, cubic foot, and cubic mile. They may also suggest measures that are not the cube of a linear unit, such as quart and gallon.

The main point of today's activity is to get students to interpret the task of finding volume as essentially counting cubes. The more complex solid figures are included to offer a challenge for students who already understand the concept.

Some of the drawings in the student book are ambiguous. If students are confused, tell them to do their best to figure out what the drawing represents and then work with whatever decision they make. The focus here is on finding volume and surface area, not on interpreting three-dimensional drawings.

Show students the cubes they will use. Clarify that they should use a single cube as the unit of volume and a face as the unit of area. Then let them begin the activity. If any general difficulties arise, you might discuss them with the whole class.

The purpose of asking for shortcuts in Question 2b is to bring out the role of multiplication in computing volumes. This idea will be addressed more systematically in *The Ins and Outs of Boxes*.

Discussing and Debriefing the Activity

No discussion of this activity is necessary, beyond perhaps a simple sharing of ideas.

Key Question

What does *volume* mean? How is it different from surface area?

Put Your Fist Into It

Intent

This activity gives students experience with a nonstandard unit for measuring volume, similar to their work in *Approximating Area.* In addition, students tackle a problem that pulls together several important ideas raised in recent activities.

Mathematics

Part I reinforces the idea that finding volume means counting how many times a fixed unit of volume can fit inside an object. It also helps students clarify criteria for a useful unit of measurement. Part II poses a task that will require students to combine their knowledge of trigonometry and the Pythagorean theorem.

Progression

Students work on the two parts of this activity individually. In Part I, they use their own fists to measure volumes of everyday objects. In Part II, they combine trigonometry and the Pythagorean theorem to analyze a diagram. They share discoveries in a class discussion.

Approximate Time

25 minutes for activity (at home or in class)
10 minutes for discussion

Classroom Organization

Individuals, followed by whole-class discussion

Doing the Activity

This activity requires little or no introduction.

Discussing and Debriefing the Activity

Ask for volunteers to describe the objects they measured and how they used their fists to find the volumes. Other volunteers might share how they estimated the number of cubic inches in their fists.

Next, have one or two presentations on Question 3, focusing on how the results in Questions 1 and 2 were combined. Students should realize, for example, that if a fist has a volume of 10 cubic inches and an object has a volume of 20 fists, then the volume of the object is 200 cubic inches.

Ask the class, Is your fist a good unit for volume? Why or why not? Bring out at least these two reasons why it is not.
- A person's fist is not a standard unit.
- Fists don't necessarily fit together well.

IMP Year 2, Do Bees Build It Best? Unit, Teacher's Guide

© 2010 Interactive Mathematics Program

112

The main point of Part II is for students to acknowledge that they sometimes need to combine ideas. They will probably use trigonometry to find lengths *BE* and *CE* and the Pythagorean theorem to find length *BC*.

Key Question

Is your fist a good unit for volume? Why or why not?

The Ins and Outs of Boxes

Intent
Students develop formulas for the surface area and volume of a rectangular solid.

Mathematics
This activity offers students practice in recognizing patterns as they discover formulas for the surface area and volume of a rectangular solid.

Progression
Students work in groups, using data from *Flat Boxes* to create In-Out tables for surface area and volume. They find rules for their tables, thus obtaining formulas for the surface area and volume of a rectangular solid. They share their formulas in a class discussion.

Approximate Time
30 minutes

Classroom Organization
Groups, followed by whole-class discussion

Doing the Activity
Ask students to think back to their work on *Flat Boxes.* If they still have the nets or boxes they created, they should take them out. Ask, What did you need to know about each box to make a net for it? The purpose of this question is to bring out that a box is "defined by" its dimensions. If you know the length, width, and height of a box (even if you don't know which is which), you know enough to construct it.

Have students read through this activity. Clarify, as needed, that surface area and volume are the *Outs* for the two In-Out tables. Each table has three *Ins,* which are the dimensions of the box.

Discussing and Debriefing the Activity
When most groups are finished, have representatives from several groups share their formulas. There probably won't be disagreement about the formula for volume, but students may find different expressions for surface area, such as $2lw + 2lh + 2wh$ and $2(lw + lh + wh)$. If so, have the class establish that the expressions are algebraically equivalent.

Key Question
What did you need to know about each box to make a net for it?

IMP Year 2, Do Bees Build It Best? Unit, Teacher's Guide

114

© 2010 Interactive Mathematics Program

Supplemental Activities

Finding the Best Box (reinforcement) follows up on this work with surface area and volume. Students can solve each example of this traditional optimization problem by trial and error and should be able to perceive the pattern.

Another Best Box (extension) involves volume and surface area and requires some algebraic manipulation.

IMP Year 2, Do Bees Build It Best? Unit, Teacher's Guide

© 2010 Interactive Mathematics Program

115

A Sculpture Garden

Intent

Students examine how to minimize surface area for a certain type of solid figure with a given volume.

Mathematics

In this activity students consider the question, "Is it possible for two figures with the same volume to have different surface areas?" They are asked to arrange eight cubes into three-dimensional shapes to determine which has the fewest faces exposed (including the bottom). For example, the figure at left has 24 faces exposed; the figure at right has 34 faces exposed. Students will learn how to draw such figures, and they will relate this question to the unit problem.

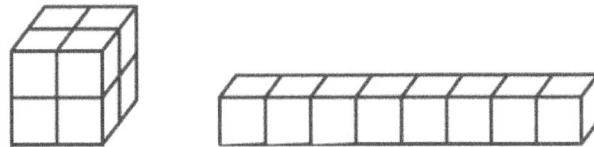

Progression

The activity is introduced through a brief discussion of techniques for drawing a cube. Students then work in pairs, first with cubes and then by sketching, to arrange eight cubes so that the resulting shapes minimize surface area. Students share findings in a class discussion.

Approximate Time

35 minutes

Classroom Organization

Pairs, followed by whole-class discussion

Materials

Multilink cubes or other cubes, if linking cubes are unavailable (8 per pair of students)

Doing the Activity

Students will find it helpful to be able to draw cubes for this activity, so you may want to take a few minutes to demonstrate ways to do this. Perhaps a couple of students who feel comfortable with three-dimensional drawing can share their techniques. Here are two common methods.

IMP Year 2, Do Bees Build It Best? Unit, Teacher's Guide

116

© 2010 Interactive Mathematics Program

Method A

Draw a square, draw an identical square slightly up and to the right of the first, and connect the four sets of corners. After students get the basic idea, they might use dashed lines to represent edges that are not visible from the front.

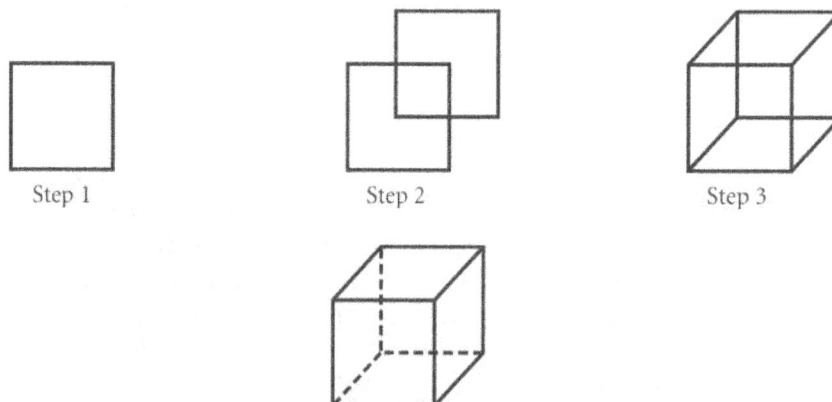

Step 1   Step 2   Step 3

Method B

Draw a square, draw three segments of equal length at about 45° angles from three of the corners, and connect the ends of these segments. In this method, only edges that are visible from the front are drawn.

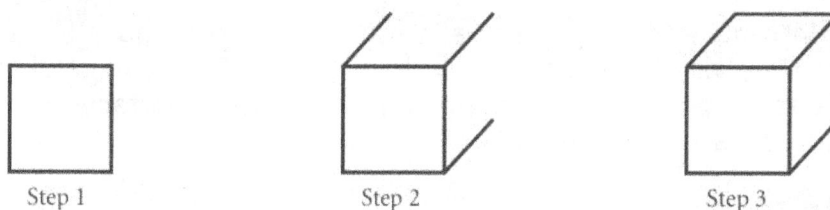

Step 1   Step 2   Step 3

Have students work in pairs, using cubes to build the economical arrangements for Neville's situation. Each pair should try to agree on the best solution, or the arrangement that will give the minimal surface area.

Discussing and Debriefing the Activity

Have pairs share their solutions. (Placing the eight cubes in a 2–by–2–by–2 arrangement is the only way to get the minimal surface area of 24 square units.)

Then ask, **What does this activity have to do with the honeycomb problem?** Try to draw out that this situation involves finding a shape with the least surface area for a given volume (8 cubic units). Also bring out that students were restricted to shapes that could be made from eight unit cubes and couldn't consider other solids with that volume.

Make sure students recognize that just as figures with the same area can have different perimeters, solids with the same volume can have different surface areas.

Key Question

What does this activity have to do with the honeycomb problem?

Reference: The World of Prisms

Intent

These reference pages give students definitions of and terminology for different types of prisms.

Mathematics

This reference material defines the terms **prism**, **base**, **lateral face**, and **lateral surface area** and introduces the classification of prisms according to the shape of the base and as **right prisms** or *oblique prisms.*

Progression

This material is intended for student reference and does not include an activity. Most of this will be new material for students, so discuss it with them and encourage them to take notes on the new vocabulary.

Approximate Time

15 minutes

Classroom Organization

Whole class

Materials

Prism models will be useful for display. A Toblerone® candy bar box is a great example.

Presenting the Reference Material

In the activity *Shedding Light on Prisms,* students will extend their earlier work with surface area and volume of rectangular solids to prisms more generally. The introductory material on prisms given here will be new to many students, but you should look for opportunities to build on their existing knowledge.

Ask students if they know what a **prism** is. They may know it as an optical instrument or a decoration. Tell them, if no one knows, that a prism is also a kind of a **polyhedron**, a many-faced solid figure. (The term *polyhedron* comes from Greek roots meaning "many-sided," while the word *polygon* comes from roots meaning "many-angled.") Review the terms **face**, **edge**, and **vertex**. You might ask, How many faces, edges, and vertices does a rectangular solid have?

You might suggest that students think of forming a prism in this way. Start with any polygon. Move the polygon through space for a fixed distance in a fixed direction, keeping it parallel to the plane of its original position and not turning it. The initial and final positions of the polygon form two of the faces of the prism and are called the **bases**. The bases of a prism are thus polygons of the same shape and size and are **congruent**. The bases are also parallel to each other. The perpendicular distance between the bases is called the *height* of the prism.

In each of the prisms here, the upper base is shaded. A prism is classified by the shape of its base. Thus, example A is a *triangular prism,* B is a *hexagonal prism,* and C is a *rectangular prism.*

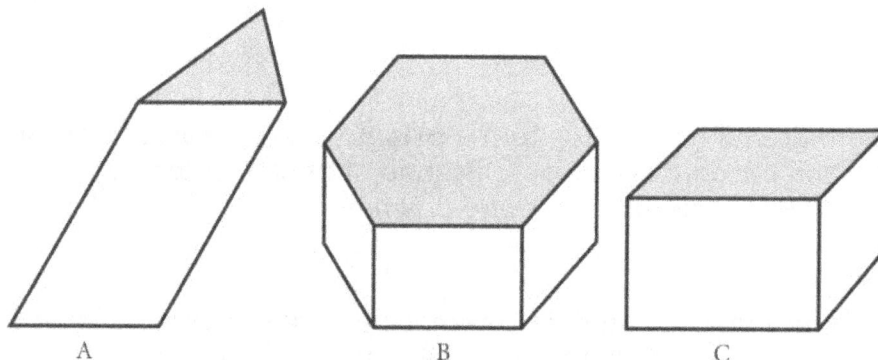

A   B   C

The other faces of a prism are called **lateral faces** (the word *lateral* means "side"), and they are all parallelograms. Each is formed by a pair of corresponding sides from the bases and the two segments connecting the corresponding vertices. (*Note:* The bases themselves can also be parallelograms, but they are not considered lateral faces.) The segments connecting corresponding vertices of the bases are called **lateral edges.**

Tell students that one important category of prisms consists of those for which the lateral edges are perpendicular to the bases. Such a prism is called a **right prism,** and the lateral faces of a right prism are all rectangles.

A rectangular solid, such as a cube, is called a *right rectangular prism,* because it is a right prism whose base is a rectangle.

A prism that is not a right prism is called an *oblique prism.* Examples B and C are right prisms (or are intended to be), and example A is intended to be an oblique prism. (In a two-dimensional drawing, it's sometimes hard to distinguish whether a prism is right or oblique.)

Finally, explain that the **lateral surface area** of a prism is the total area of the lateral faces, or the prism's total surface area minus the area of the two bases.

Key Question

How many faces, edges, and vertices does a rectangular solid have?

Shedding Light on Prisms

Intent

Students develop formulas for the lateral surface area and volume of a right prism. These formulas will play an important role in the unit problem.

Mathematics

The primary goal of this activity is for students to discover two formulas related to right prisms.

Volume = height • area of the base (understood by visualizing the number of volume units that cover the base and the number of layers of these units)

Lateral surface area = height • perimeter of the base (understood by visualizing the rectangle that wraps around the sides of the prism)

The activity is set up to facilitate students' reaching these conclusions through their own observations. Students may discover the formulas either by recognizing relationships among the numeric results or by a more geometric or algebraic analysis. The formulas are then generalized to arbitrary prisms as well as to cylinders.

Progression

Students build several prisms. They measure the height, the area of the base, the volume, the perimeter of the base, and the lateral surface area of each prism and then look for relationships among these measurements. A class discussion then explores why these formulas make sense and generalizes them to cylinders.

Approximate Time

40 minutes

Classroom Organization

Groups, followed by whole-class discussion

Materials

Multilink cubes or other cubes, if linking cubes are unavailable (54 per group)

Doing the Activity

Before students begin work, review the units of measurement they will use. The cubes are to be used as the basis for measuring length, area, and volume. If you have linking materials in the form of triangular prisms, you can use them to create additional problems for this activity.

If you don't observe groups developing the formulas as they work, look for opportunities to build on what they are doing with specific examples. For instance,

IMP Year 2, Do Bees Build It Best? Unit, Teacher's Guide

© 2010 Interactive Mathematics Program

120

in Question 4, students are probably not counting all the cubes; more likely, they are recognizing that there are 6 cubes in each layer. You can help them discover the volume formula listed above by asking, How do you know there are 54 cubes without counting?

Suggest students put their results from Part I into a table. (As part of their task is to figure out which parameters should be thought of as functions of which others, exercise restraint in structuring such a table for them.)

Discussing and Debriefing the Activity

The formulas that students discover in this activity are crucial to the future development of this unit, so you may want to allow extra time if groups don't seem to be finding them. If some groups have found the formulas and others are still struggling, you might have groups that are done plan presentations while the others continue work.

Have several students present their conclusions. You might have the first group present numeric data and let subsequent groups focus on the general relationships. Be sure both of these principles are clearly articulated:

Volume = height • area of the base

Lateral surface area = height • perimeter of the base

Why Do These Formulas Work?

Ask the class, How can you explain these two formulas in terms of the geometry? Many students will probably be able to explain the volume formula because they found the volumes in this activity using essentially this idea. For example, for Question 4, they probably multiplied the 9 layers of cubes by the 6 cubes in each layer to get 54 units as the volume.

Explaining the second formula will likely be more difficult. You might ask, How did you get the surface area of each face? Students probably multiplied the length of the appropriate side of the base by the height. To find the total lateral surface area, they likely added these products, which are each of the form hs, where h is the height and s is the length of a side of a base. In other words, they may in effect have used a formula like

$$\text{Lateral surface area} = hs_1 + hs_2 + \ldots + hs_n$$

where $s_1, s_2, \ldots, s_n$ are the lengths of the sides.

If students can articulate this overall process explicitly, they can then use the distributive property (either formally or intuitively) to recognize that the lateral surface area can also be found as $h(s_1 + s_2 + \ldots + s_n)$. This gives the second formula, because $s_1 + s_2 + \ldots + s_n$ is the perimeter of the base.

IMP Year 2, Do Bees Build It Best? Unit, Teacher's Guide

© 2010 Interactive Mathematics Program

121

Another useful approach is to ask, What would a net look like that could be folded to form only the sides of the prism? Students should recognize that this net would itself be a rectangle that wraps around the prism with height equal to the height of the prism and width equal to the perimeter of the base of the prism. The area of that rectangle is the lateral surface area of the prism.

Connections to a Cylinder

One way to deepen students' understanding of the concepts is to ask, Do these formulas apply if the base is something other than a polygon? In particular, raise the case of a cylinder. What is the familiar name for a solid figure with a circular base? Students should recognize that this solid figure is a cylinder. They should also be able to generalize from their work that the volume of a cylinder is the product of the height and the area of the circular base and that the lateral surface area is the product of the height and the circumference of the circular base.

Key Questions

How do you know there are 54 cubes without counting? How can you explain these two formulas in terms of the geometry?

How did you get the surface area of each face?

What would a net look like that could be folded to form only the sides of the prism?

Do these formulas apply if the base is something other than a polygon?

What is the familiar name for a solid figure with a circular base?

Pythagoras and the Box

Intent

Students apply the Pythagorean theorem in a three-dimensional context.

Mathematics

Students are asked to determine whether an object of a given length will fit into a particular box and then to develop a formula for finding the length of the diagonal of a box. This work generalizes the Pythagorean theorem to three dimensions.

Progression

Students work on the three questions in this activity on their own and discuss their results as a class.

Approximate Time

25 minutes for activity (at home or in class)
10 minutes for discussion

Classroom Organization

Individuals, followed by whole-class discussion

Doing the Activity

You might want to use a classroom object (or perhaps the classroom itself) to clarify the situation.

Discussing and Debriefing the Activity

Students who easily visualize situations in three dimensions have an advantage in this activity. You may want to give students time to discuss their results in groups before asking for a volunteer to make a presentation on Question 1. Be sure the presenter demonstrates how to find the actual length of the diagonal. Students should have found that the pen will not fit because the length of the diagonal is only $\sqrt{84}$, or 9.17, inches.

Have one or two volunteers present their examples from Question 2. Then let a volunteer present ideas for the general formula (Question 3). You might suggest that students think of this formula as a "three-dimensional Pythagorean theorem."

IMP Year 2, Do Bees Build It Best? Unit, Teacher's Guide

123

Back on the Farm

Intent

Students find the volumes and lateral surface areas of several prisms and have an opportunity to apply what they learned in *Shedding Light on Prisms.*

Mathematics

The two situations in this activity deal with volume and lateral surface area for right prisms with bases that are regular polygons with other than four sides. Working with the prism in a horizontal orientation in Question 1 will encourage students to recognize that orientation does not determine which faces of a prism constitute the bases. In Question 2, they are reminded that if the perimeter of the base of a prism is fixed, the number of lateral faces does not affect the prism's lateral surface area.

Progression

Working individually, students are asked to find the volume of a right triangular prism that is not sitting on one of its bases. Then they consider how changing the number of sides affects the lateral surface area if the perimeter of the base is held constant. They share discoveries in a class discussion.

Approximate Time

20 minutes for activity (at home or in class)
15 minutes for discussion

Classroom Organization

Individuals, followed by whole-class discussion

Doing the Activity

In Question 1, the prism is not in the typical position with the base horizontal. If students find this confusing, you might suggest they visualize the trough stood on its end.

Discussing and Debriefing the Activity

Ask for a volunteer to explain the answer to Question 1. Here are two ways to approach this question.

- Students can apply the general principle that the volume of a right prism is the product of its height and the area of its base. To do this, they find the area of the triangular base (0.5 square foot) and multiply it by the length of the trough (5 feet) to get a volume of 2.5 cubic feet.
- Students can observe that if a second, identical trough is placed upside down on top of the first, the two make a rectangular solid 1 foot by 1 foot by 5 feet, for a volume of 5 cubic feet. Thus one trough has a volume of 2.5 cubic feet.

Ask for volunteers to talk about results for Question 2. Here are two of the various approaches students might use.

- Students can apply what they learned in *Shedding Light on Prisms,* multiplying the perimeter of the base by the height of the prism.
- Students can do a face-by-face analysis, adding the separate areas.

Those who used the second approach probably examined the three cases (hexagon, decagon, and dodecagon) separately, or at least two of them, before realizing they all give the same result.

If the first presenter uses the second approach—and you might consider orchestrating the discussion so this happens—ask whether anyone can explain why the answers are the same. If needed, refer students to their work on *Shedding Light on Prisms* or lead them through an algebraic explanation based on the distributive property as described in the discussion of that activity.

IMP Year 2, Do Bees Build It Best? Unit, Teacher's Guide

125

Which Holds More?

Intent

This activity reinforces the concepts of volume and surface area by extending them to the context of a cylinder.

Mathematics

A standard-size sheet of paper can be used to create two different cylinders. These cylinders have the same surface area: the area of the sheet of paper. However, their volumes are different. This activity offers students more exploration of the ideas behind the formulas for the volume and surface area of solids, this time using a cylinder.

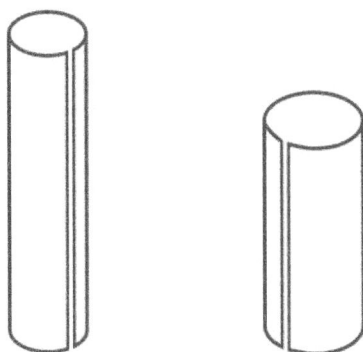

Progression

Students work in groups to compare the volume and lateral surface area of two cylinders formed from a sheet of paper. The class then discusses the findings.

Approximate Time

35 minutes

Classroom Organization

Groups, followed by whole-class discussion

Materials

Puffed rice, dried beans, or similar material for measuring volume
8.5-by-11-inch sheets of paper

Doing the Activity

This activity requires little or no introduction.

Discussing and Debriefing the Activity

When you discuss Question 1, keep in mind that students have not yet encountered the formula for area of a circle in the IMP curriculum. Their work on this question will thus probably be intuitive.

IMP Year 2, Do Bees Build It Best? Unit, Teacher's Guide

126

© 2010 Interactive Mathematics Program

Some students will intuitively sense that the shorter, wider figure has the greater volume, and you can confirm this by filling each cylinder with puffed rice and comparing the quantities required.

For Question 2, some students will recognize that because the lateral surface area of both cylinders is just the area of the paper, the two surface areas are the same.

As an alternative approach, you might ask, **What is the circumference of the base of each cylinder?** Students should recognize that the two circumferences are $8\frac{1}{2}$ inches and 11 inches. With this information, they might find the surface areas from the principle that the lateral surface area of a cylinder is the product of the height and the circumference of the base, an idea that should have come out in the discussion of *Shedding Light on Prisms.*

Key Question
What is the circumference of the base of each cylinder?

Cereal Box Sizes

Intent

In this activity and *More About Cereal Boxes,* the goal is for students to become aware of the proportionality principles involved in finding volumes and surface areas of similar rectangular solids. Here the focus is on volume; the next activity focuses on surface area.

Mathematics

This work builds on students' experiences in *More Fencing, Bigger Corrals,* in which they found that if two similar figures have perimeters in the ratio $n:1$, their areas will be in the ratio $n^2:1$. In other words, if the dimensions of a polygon are multiplied by a scale factor, the area of the polygon will be multiplied by the square of that scale factor.

This activity extends this principle to three-dimensional similar figures. Students come to realize that if the dimensions of a polyhedron are multiplied by a scale factor, the volume is multiplied by the cube of that scale factor.

Progression

Students work on the activity individually. The follow-up discussion focuses on the generalizations requested in Questions 4b and 4c.

Approximate Time

20 minutes for activity (at home or in class)
10 minutes for discussion

Classroom Organization

Individuals, followed by whole-class discussion

Doing the Activity

This activity requires little or no introduction.

Discussing and Debriefing the Activity

Have students present Questions 1 and 2. You may also want to have volunteers discuss Question 3.

Students will likely have recognized that the volume of the minisize box is $\frac{1}{8}$ the volume of the standard box and that the volume of the institutional-size box is 27 times that of the standard box. The goal is to have them understand this geometrically as well as numerically. For example, they should be able to describe how 8 minisize boxes can be stacked to make a standard box or how 27 standard boxes can be put together (in a 3-by-3-by-3 arrangement) to make an institutional-size box.

Students should also be able to state that the volume of the supersize box is 125 times that of the standard box and to explain that this is because 5^3 is 125. You might build on their experience with Emma and her crates in *A Sculpture Garden* to help them visualize how smaller boxes can be put together to make a larger box.

Encourage students to articulate the general principle involved. They might state it like this.

If all the dimensions of a rectangular solid are multiplied by the same factor, then the volume is multiplied by the cube of that factor.

Ask students, *How would you generalize this principle to any pair of similar solids (not just boxes)?* You might focus their attention on the ratio of the volumes of such figures. The goal of the discussion is to elicit a general principle like this one.

If two solid figures are similar, then the ratio of their volumes is the cube of the ratio of corresponding linear dimensions.

Strictly speaking, we have only defined the concept of similarity for polygons, not for solids, but students should be able to apply their intuitive notion of "same shape" to extend the concept.

Key Question

How would you generalize this principle to any pair of similar solids (not just boxes)?

IMP Year 2, Do Bees Build It Best? Unit, Teacher's Guide

129

© 2010 Interactive Mathematics Program

More About Cereal Boxes

Intent

Students continue their work from the previous activity, *Cereal Box Sizes,* now focusing on surface area.

Mathematics

This activity leads students to the discovery of the principle that if the dimensions of a polyhedron are multiplied by a scale factor, the surface area is multiplied by the square of that scale factor. Because the faces of a polyhedron are polygons, this principle follows directly from students' work on *More Fencing, Bigger Corrals.*

Progression

Students work on the activity individually or in groups. The follow-up discussion focuses on the general principles uncovered in Question 3.

Approximate Time

40 minutes

Classroom Organization

Individuals or groups, followed by whole-class discussion

Doing the Activity

This activity requires little or no introduction.

Discussing and Debriefing the Activity

Similar to what they did for volume in *Cereal Box Sizes,* students should be able to perform an analysis for surface area. The surface area of the minisize box is exactly $\frac{1}{4}$ that of the standard box, the surface area of the institutional-size box is exactly 9 times that of the standard box, and the surface area of the supersize box is exactly 25 times that of the standard box. Get them to articulate the general principle involved. They may state it similarly to the principle for volume.

If all the dimensions of a rectangular solid are multiplied by the same factor, then the surface area is multiplied by the square of that factor.

Ask how this principle relates to the one established in the discussion of *More Fencing, Bigger Corrals.* That principle, which may have been posted previously, was something like this.

If two polygons are similar, then the ratio of their areas is the square of the ratio of their perimeters.

Help students to realize that the new statement is a particular case of the earlier principle. Bring out that when the dimensions of a rectangular solid are all

IMP Year 2, Do Bees Build It Best? Unit, *Teacher's Guide*

© 2010 Interactive Mathematics Program

130

multiplied by the same factor, corresponding pairs of faces of the smaller and larger solids are similar rectangles.

You may also want to have students compare their conclusions from *Cereal Box Sizes* and this activity. They might note these connections.

- Volume is a three-dimensional measurement, measured in such units as cubic inches, so the volumes of similar figures are in the ratio of the *cube* of the ratio of lengths.
- Surface area is a two-dimensional measurement, measured in such units as square inches, so the surface areas of similar figures are in the ratio of the *square* of the ratio of lengths.

IMP Year 2, Do Bees Build It Best? Unit, Teacher's Guide

131

A Size Summary

Intent

This activity follows up students' work in *Cereal Box Sizes* and *More Cereal Boxes,* reinforcing the principles of proportionality for volume and surface area of similar solids.

Mathematics

Students practice finding the volume and surface area of rectangular prisms and applying the principles that the surface area and volume of similar solids are proportional to the square and the cube of the scale factor, respectively.

Progression

Students work individually to summarize and explain their conclusions from the two "cereal box" activities and share them in a class discussion.

Approximate Time

20 minutes for activity (at home or in class)
10 minutes for discussion

Classroom Organization

Individuals, followed by whole-class discussion

Doing the Activity

This activity requires little or no introduction.

Discussing and Debriefing the Activity

You might begin by having students state the general principles, while focusing on the specific examples.

Question 2a reinforces ideas from *The Ins and Outs of Boxes,* and you can clear up any difficulties students are having.

For Question 2b, students should calculate that the volume and surface area are multiplied, respectively, by 3^3 and 3^2. Be sure they articulate this relationship, applying the principles of proportionality, rather than just giving the new volume and surface area by direct computation. How can you find the answer to Question 2b without multiplying 6 by 9 by 15?

For Question 2c, the goal is for students to realize that they want the volume to be 5 times the original. That means finding a scale factor whose cube is 5. (You might introduce the term *cube root* if students don't use it.) Because the cube root of 5 is about 1.709, the desired rectangular solid is roughly 3.4 inches by 5.1 inches by 8.5 inches. Note that rounding $\sqrt[3]{5}$ to different degrees of precision before

IMP Year 2, Do Bees Build It Best? Unit, Teacher's Guide

132

© 2010 Interactive Mathematics Program

multiplying affects the results, as in *Falling Bridges.* To test the result, students can verify that the product of these numbers is roughly 150.

Similarly, for Question 2d, students should notice that the surface area has been doubled, which means they want to multiply the dimensions by $\sqrt{2}$, or approximately 1.414. (Again, rounding this to different places produces different results.) This gives dimensions of 2.8 inches by 4.2 inches by 7.1 inches. Again, students can verify that this gives the correct surface area.

Key Question

How can you find the answer to Question 2b without multiplying 6 by 9 by 15?

Back to the Bees

Intent

These activities, which begin with a teacher-led discussion, synthesize students' earlier work and answer the unit question.

Mathematics

At the start of the unit, the unit question was somewhat generally stated as, "Why do honeycombs look the way they do?" The question is now made more precise and more mathematical, based on the work done throughout the unit, as, "Which right prism maximizes the volume for a given surface area and tessellates?"

Progression

Students review their work from the unit, answer the "mathematized" unit question, and compile their unit portfolios. They also take the unit assessments.

Bees Summary Discussion

A-Tessellating We Go

A Portfolio of Formulas

Do Bees Build It Best? Portfolio

Bees Summary Discussion

Intent

In this teacher-led discussion, students return to the unit problem and restate it using the ideas developed throughout the unit.

Mathematics

All the major topics of this unit—calculating surface area and volume, maximizing area in a polygon of a given perimeter, and tessellation—are synthesized in this discussion for the purposes of redefining the unit problem. With the simplifying assumptions that "a given amount of wax" means that the surface area of the honeycomb cell is fixed, and that the height of the cell is also fixed, solving the unit problem reduces to finding the base of a right prism that (1) maximizes its volume for a fixed lateral surface area and (2) tessellates.

Progression

This teacher-led discussion reviews and pulls together students' work throughout this unit and looks at how it relates to the unit problem. Students examine an actual honeycomb and review their work on several key problems from the unit.

Approximate Time

15 minutes

Classroom Organization

Whole-class discussion with group work

Materials

Honeycomb or a large photo of a honeycomb

Discussing and Debriefing the Activity

Introduction to the Discussion

The given amount of wax in the unit problem can be considered as a given surface area for the honeycomb cell. New simplifications are introduced by deciding that (1) we will consider only the lateral surface area as we analyze the efficiency of the cell with respect to volume for a given surface area (amount of wax) and (2) we will assume the height of the cells is fixed.

In this activity students combine two ideas—the principle that maximizing the number of sides in a polygon of a given perimeter maximizes the polygon's area, and their understanding of how to calculate the volume and surface area of a prism. This combination dictates that the volume of the honeycomb cell can be maximized by a base that has as many sides as possible.

IMP Year 2, Do Bees Build It Best? Unit, Teacher's Guide

135

© 2010 Interactive Mathematics Program

In other words, among right prisms of a given height and lateral surface area whose bases are regular polygons, the more sides the polygonal base has, the greater the volume of the prism. This will lead to the conclusion that a cylinder would be the most efficient shape for the base.

Further consideration will reveal that the material can be used more efficiently if the shape for the base tessellates.

Conducting the Discussion

You may want to pass around a honeycomb. Ask someone to remind the class of the question they are trying to answer about bees. "What shape honeycomb cell will give the most volume for a given amount of wax?"

Then ask, **What type of measurement should we use to talk about "amount of wax"?** As discussed earlier (recall, for example, the remarks made in introducing *Flat Cubes*), surface area will be used to evaluate the amount of wax in a honeycomb cell. It may help to remind students again of their work building containers from construction paper to justify using surface area in a situation that involves volume.

What do prisms have to do with the honeycomb question? You might return to students' observations about containers in *What to Put It In?* They probably observed that most containers are either rectangular solids or cylinders. You can ask what these two solids have in common. Bring out that the rectangular solids are right prisms and that the cylinders are similar but with one curved side instead of several flat ones. Tell the class that from now on, they should limit their candidates for the honeycomb cells to prism-like objects, either right prisms or right circular cylinders.

Choosing the Shape of the Base

Choosing a prism-like object can be thought of as answering these two questions.
- What should the height of the object be?
- What should the shape of the object's base be?

This unit focuses on the second of these questions and makes the additional simplification of looking only at *lateral* surface area. This approach allows students to apply their work in *Shedding Light on Prisms* and considerably simplifies the problem, essentially reducing it to something equivalent to the corral problem.

Even with this simplification (considering only lateral surface area), it is important to fix the height. As students realized in *Which Holds More?*, different volumes can be obtained from the same lateral surface area—with the same base shape—by changing the balance between height and circumference of the base.

So, tell students that to make the problem more manageable, they should make two other simplifications.
- They should consider only the cell's lateral surface area rather than its total surface area.

- They should assume that the height of the cell has been fixed and that they only need to decide on the shape of the base.

From now on, students can thus assume they have a fixed amount of wax for the lateral surface of the honeycomb and that the height of the cell has also been fixed. Ask students to discuss in their groups how they might use their work with prisms to decide on a shape for the base of the cell.

You may want to remind students of the formulas they developed in *Shedding Light on Prisms,* emphasizing that both volume and surface area are computed using the height as a factor. They need to put these two facts together:
- Because the lateral surface area and the height of the cell are fixed, the perimeter of the base is also determined.
- Because the cell's height is fixed, the cell's volume is maximized by maximizing the area of the base.

In other words, the problem of choosing a shape for the base comes down to this question. For a fixed perimeter, what base shape will give the greatest area, and thus the greatest volume, for a honeycomb cell?

Maximizing Area for a Fixed Perimeter
Students may recognize this question as being essentially the same one raised following the discussion of *Building the Best Fence.* If not, ask groups to discuss the question of maximizing area for a given perimeter. Students should recall that regular polygons seem to be the most efficient and that the more sides a regular polygon has (for a fixed perimeter), the greater the area. In fact, they decided that the most efficient shape seems to be a circle.

Ask, What does this say about the shape of the honeycomb cell? Students should recognize that for a given lateral surface area, a cylinder will have the most volume. You might mention that this is one reason why so many containers in stores are cylindrical.

Why Not Cylinders?
Follow the last conclusion by pointing out that honeycomb cells are not cylindrical (though they are close). Then ask, Why don't bees use cylinders? Two or more good answers might emerge. One is that circles may be hard to make. The other is that cylinders don't tessellate; they don't fit neatly together.

Point out that if the bees use a prism with a base shape that tessellates, each prism wall will serve as the wall of two honeycomb cells. Students thus need to restate the question something like this.

What shape for the base of a prism honeycomb cell, *among those that tessellate,* will give the greatest area for a base with a given perimeter?

Explain that as with the corral problems, students should consider only regular polygons for the base. They will recall that for rectangles and triangles, regular polygons are the most efficient.

Key Questions

What type of measurement should we use to talk about "amount of wax"?

What do prisms have to do with the honeycomb question?

For a fixed perimeter, what base shape will give the greatest area for a prism honeycomb cell?

Supplemental Activity

From Polygons to Prisms (reinforcement) continues the work on the connection between the shape of the base of a prism and the prism's volume.

A-Tessellating We Go

Intent

Students determine which regular polygons tessellate.

Mathematics

In the POW *Tessellation Pictures,* students played with tessellations created by figures made by transforming rectangles. Now they are asked to determine which regular polygons tessellate. They will learn that because only polygons with an angle that is a factor of 360° will work, there are only three such regular polygons: equilateral triangles, squares, and regular hexagons.

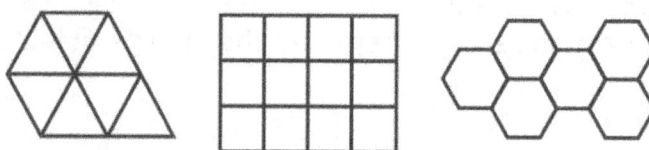

Progression

Students work in groups to determine which regular polygons tessellate and to try to explain why their conclusions are true. They share their conclusions in a class discussion.

Approximate Time

25 minutes

Classroom Organization

Groups, followed by whole-class discussion

Materials

Pattern blocks
A-Tessellating We Go blackline master (2 sheets per group)

Doing the Activity

Groups can begin by verifying that all the regular polygons in the pattern block set tessellate. They can then work with the pattern blocks, polygons cut from the "A-Tessellating We Go" blackline master, and other shapes they cut out.

As groups begin to draw conclusions from their experiments, urge them to look for an explanation of their results. The key steps in the analysis are (1) to picture several copies of a given regular polygon fitting together around a single point and (2) to realize that the angles of a regular polygon are equal and that the sum of the angles around a point must add to 360°. These two facts reveal that a regular polygon will tessellate only if the size of its angles is a factor of 360°.

Students will then need to develop a method for finding the angles of regular polygons, at least enough of a method to determine which have angles that evenly divide 360°. After they work on their own for a while, you might want to start pointing them toward these discoveries.

Look at a single point where the vertices of several tessellating regular polygons all come together. What do you know about the angles surrounding that point? How can that help you decide if a shape is going to tessellate?

Discussing and Debriefing the Activity

Ask groups to report their findings. They should have discovered that the only regular polygons that tessellate are those in the pattern block set:
  • equilateral triangles (whose angles are all 60°)
  • squares (whose angles are all 90°)
  • regular hexagons (whose angles are all 120°)

Let students state the reason why regular pentagons and regular polygons with more than six sides won't tessellate. Although they can find this experimentally, they may also be able to work out the reasoning based on the angles. Each angle of a regular pentagon is 108°, so pentagons don't tessellate. Regular polygons with more than six sides have angles between 120° and 180°. Because no number between 120 and 180 is a factor of 360, no polygon with more than six sides tessellates.

Ask what this result means for the unit problem. Students may be ready to conjecture that within the restrictions and simplifications imposed in this unit, the best shape for the honeycomb cell is a right prism whose base is a regular hexagon. In the discussion following *A Portfolio of Formulas,* they will have another opportunity to put all the ideas of the unit together.

Key Question

Look at a single point where the vertices of several tessellating regular polygons all come together. What do you know about the angles surrounding that point? How can that help you decide if a shape is going to tessellate?

Supplemental Activity

What Else Tessellates? (extension) broadens the concept of tessellation to nonregular polygons, especially nonregular triangles and quadrilaterals.

A Portfolio of Formulas

Intent
Students begin the process of reviewing their work over the entire unit, in anticipation of compiling their unit portfolios.

Mathematics
Students review these formulas developed throughout the unit:
- area of rectangles, triangles, parallelograms, and trapezoids
- the Pythagorean theorem
- the basic trigonometric ratios
- the surface area and volume of a prism and a cylinder

Progression
Students work independently to compile all the formulas from the unit and to write an explanation of each, including definitions for the variables. In the follow-up discussion, the various principles from the unit are brought together to solve the unit problem.

Approximate Time
45 minutes for activity (at home or in class)
30–45 minutes for discussion and optional video

Classroom Organization
Individuals, then groups, followed by whole-class discussion

Materials
Video from the Internet or other source (optional)

Doing the Activity
Tell students that this activity will be a required part of their portfolios.

Discussing and Debriefing the Activity
Give students a few minutes to share ideas and formulas in their groups. You may then want to bring the class together to compile a list of formulas.

The unit question (subject to several simplifying assumptions) has now been answered. You may wish to review the basic steps.
- Examining areas of polygons with a fixed perimeter and concluding that at least for regular polygons, the more sides, the greater the area
- Developing the general principle that for right prisms, maximizing volume for a fixed lateral surface area and height is the same as maximizing the area of the base for a fixed perimeter
- Limiting consideration to right prisms whose bases are regular polygons that tessellate and recognizing that this means considering only equilateral triangles, squares, and regular hexagons

IMP Year 2, Do Bees Build It Best? Unit, *Teacher's Guide*

141

This is the climax to a long, multifaceted investigation, so play it up appropriately. Students should conclude that

If one considers only right prisms whose bases are regular polygons that tessellate, then the prism with a hexagonal base is the best possible shape for a honeycomb cell.

Of course, "best possible shape" means the shape that gives the maximum volume for a fixed lateral surface area.

Optional: Video on Bees or Other Recapitulation
This is a good time to verify the solution to the honeycomb problem by showing a nature video on the topic, by inviting a beekeeper or a biologist to meet the class, or by using other community resources. Students' research from the beginning of the unit might provide sources for this review.

You might bring out that although hexagonal prisms are the most efficient shape within the constraints imposed in this unit, a slight improvement in the volume-to-surface-area ratio is possible if one allows polyhedra other than prisms. However, the most efficient shape is much more complicated and saves only about 1% on the surface area to achieve the same volume as the hexagonal prism.

Do Bees Build It Best? Portfolio

Intent

In this final activity, students compile their unit portfolios. The IMP portfolio, a documentation of what a student knows and can do, serves two primary purposes. First, the process of looking back through their work to summarize the main ideas of a unit and to identify what they learned well (and not so well) gives students an opportunity to reflect on their learning, both as a mathematical review and to recognize their accomplishments. Second, the portfolio provides an assessment opportunity for teachers, to find out what students identify as most valuable to their learning, addressing both their mathematical and nonmathematical growth.

Mathematics

The portfolio gives students an important opportunity to reflect on the many topics in this unit and how they interrelate. They will include the list of formulas they compiled in *A Portfolio of Formulas,* as well as representative samples of their work.

Progression

Students work independently to review their work over the unit, select samples, reflect on the evidence of their learning the samples demonstrate, and write cover letters. They will have the opportunity to share their portfolios with classmates.

Approximate Time

5 minutes for introduction (at home or in class)
10 minutes for discussion
45 minutes for activity (at home)

Classroom Organization

Individuals, followed by whole-class discussion

Doing the Activity

You may want to give students more than one night to complete their portfolios. Depending on what materials you allow them to use on the unit assessments, remind them to bring to class their portfolios and any other helpful work.

Discussing and Debriefing the Activity

Let volunteers share their cover letters, and then have the class brainstorm about all they have learned. It may be interesting to examine the combination of traditional and nontraditional concepts they have worked with.

Here are some of the topics they have studied.
* Area of two-dimensional figures
* The Pythagorean theorem
* Applications of trigonometry
* Volume of prisms, including rectangular solids and cylinders
* Surface area of prisms, including rectangular solids and cylinders
* Tessellations

IMP Year 2, Do Bees Build It Best? Unit, Teacher's Guide

143

Blackline Masters

Geoboard Paper

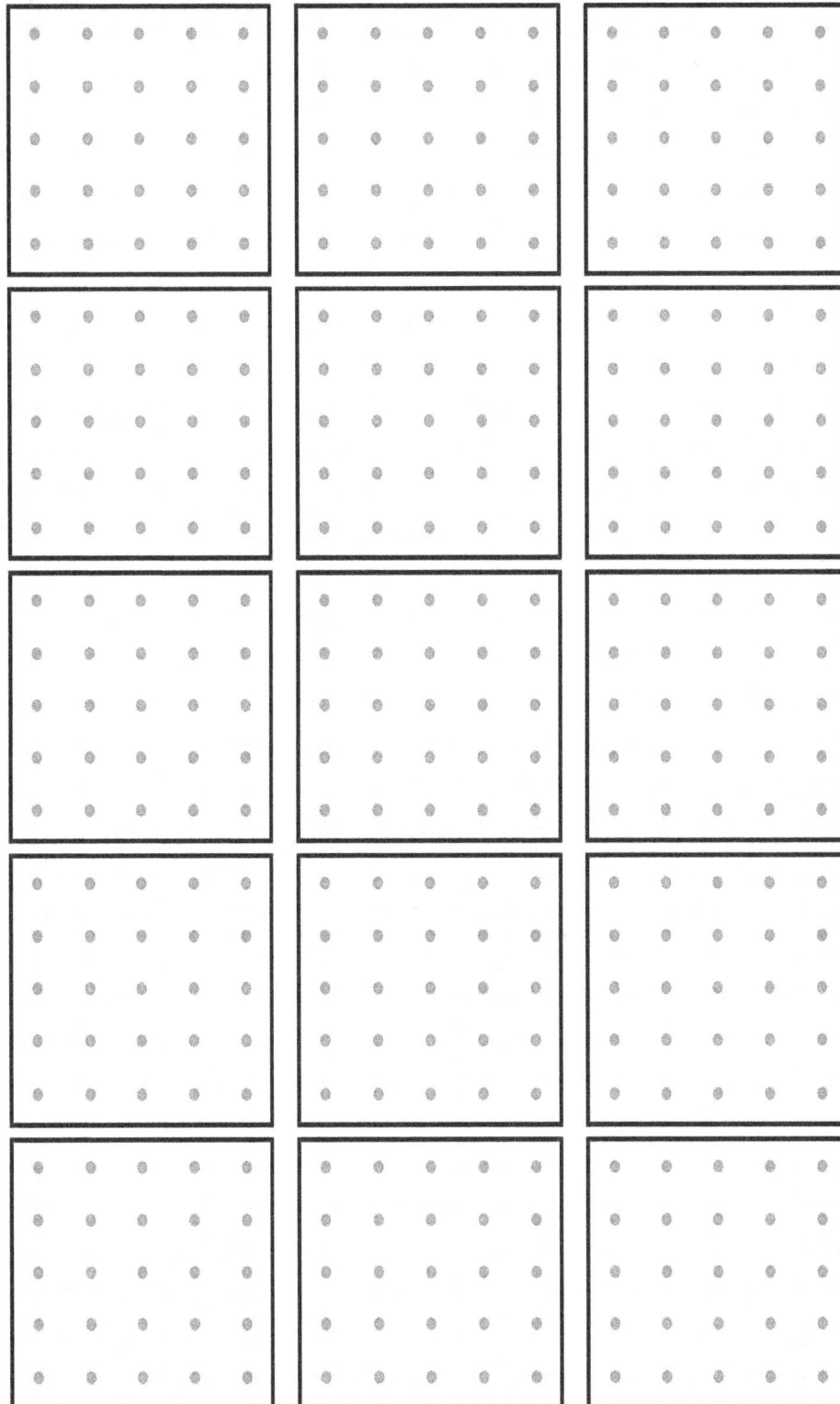

© 2010 Interactive Mathematics Program

Overhead Geoboard

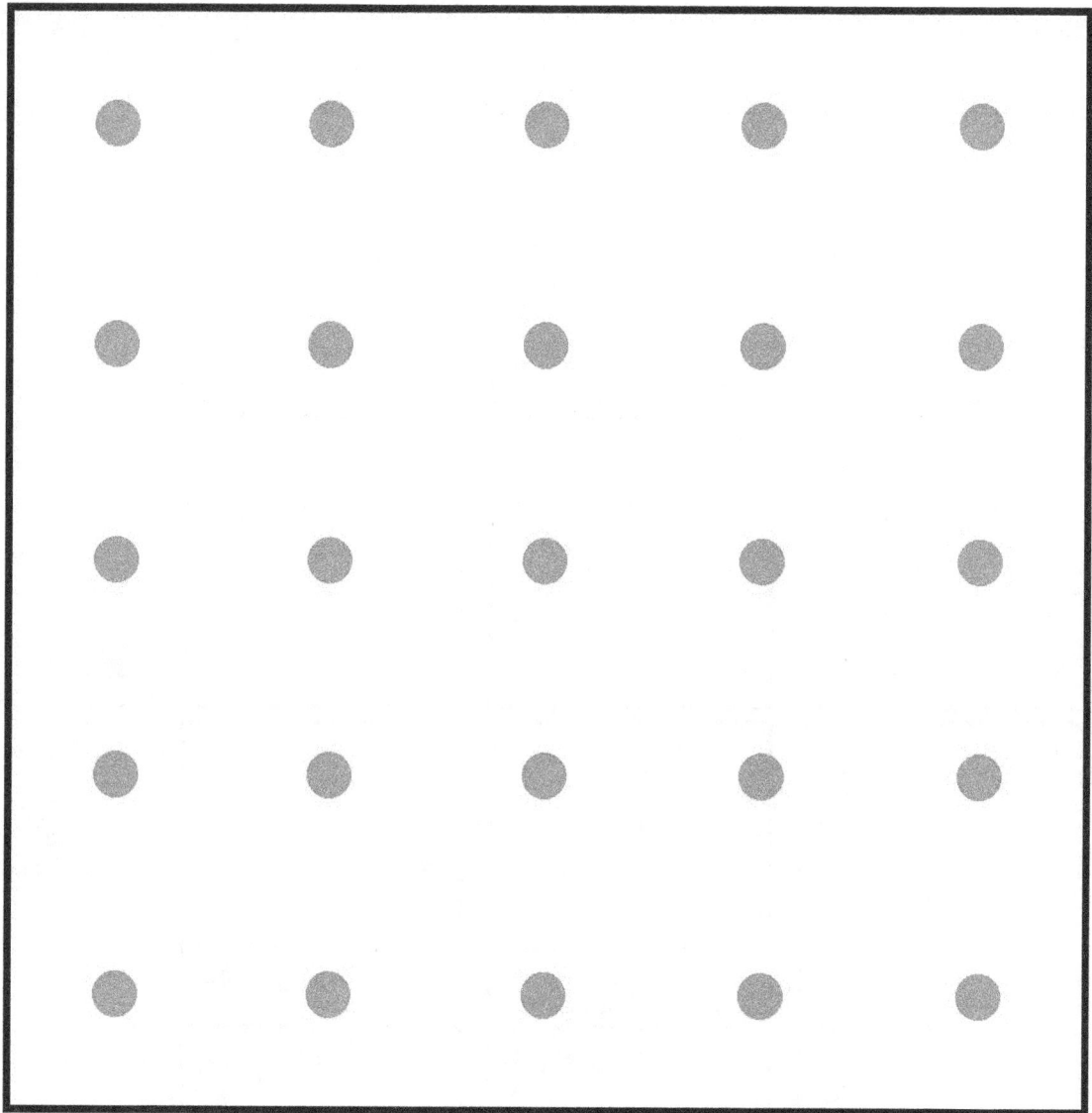

IMP Year 2, Do Bees Built It Best? Unit, Teacher's Guide

145

Halving Your Way

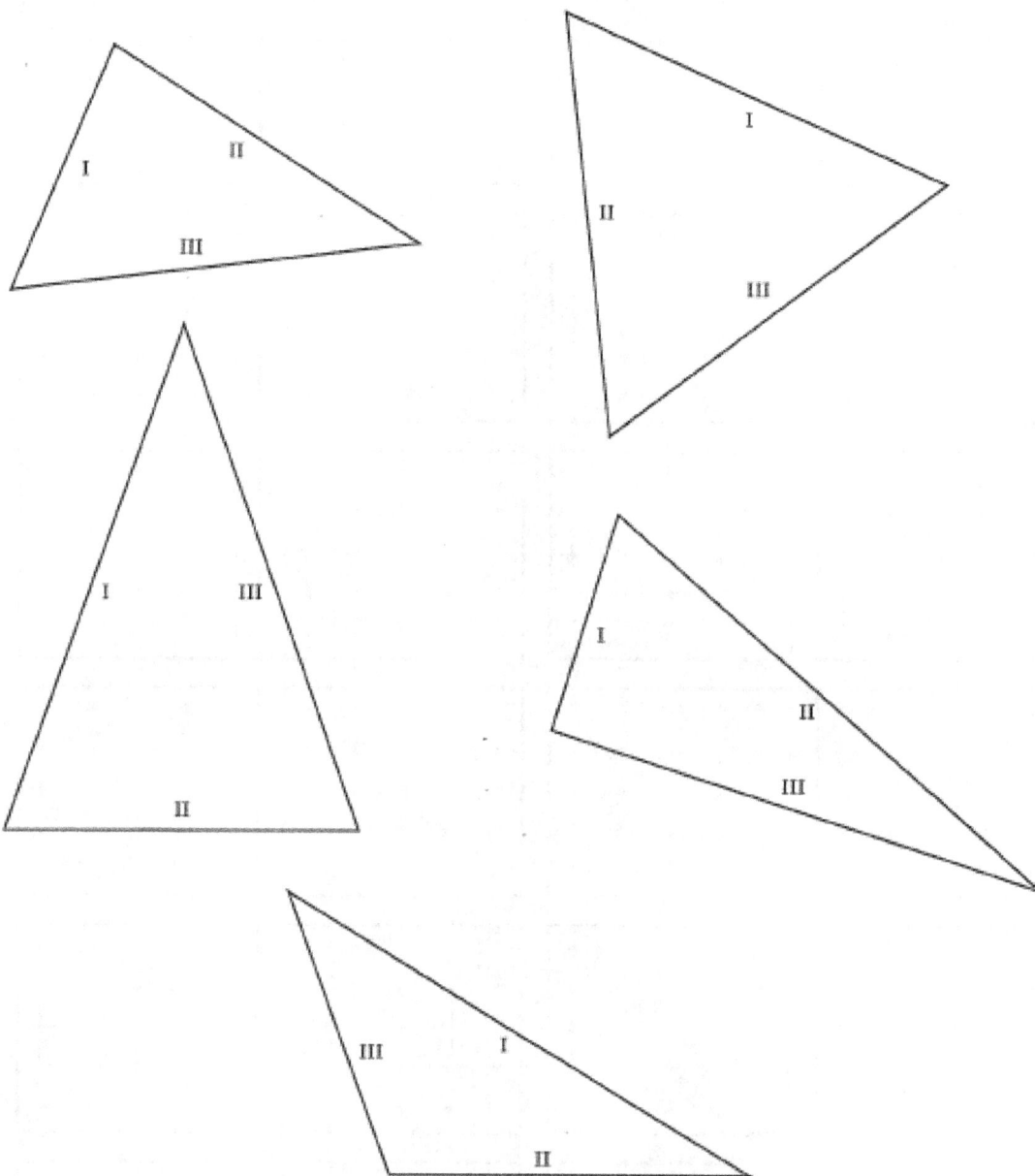

IMP Year 2, Do Bees Built It Best? Unit, Teacher's Guide

147

Metric Ruler and Protractor

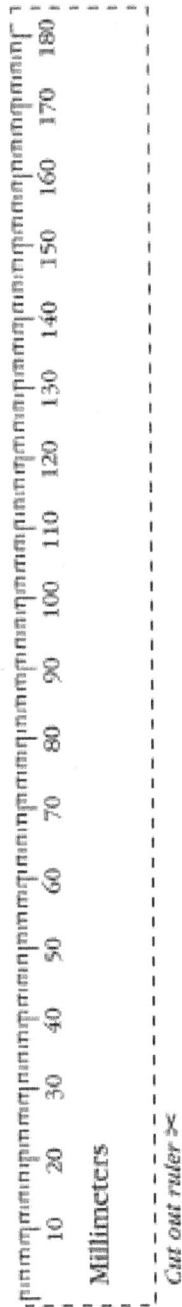

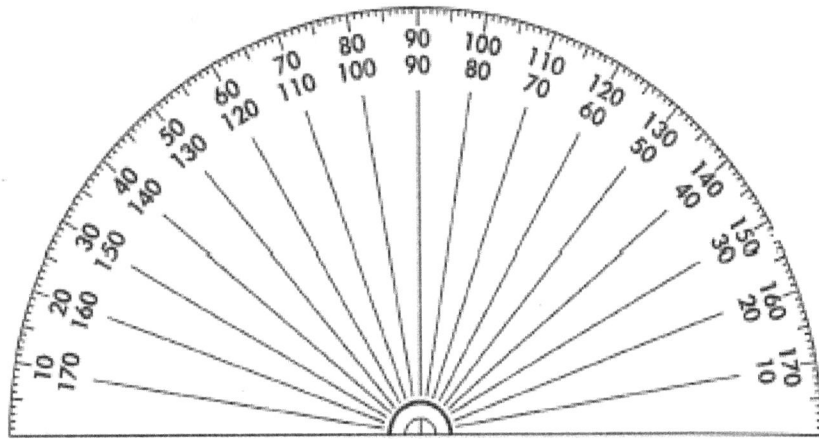

IMP Year 2, Do Bees Built It Best? Unit, Teacher's Guide

© 2010 Interactive Mathematics Program

148

 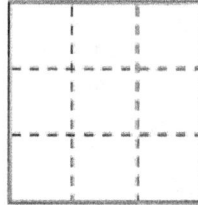 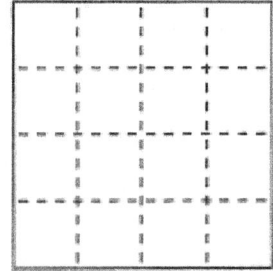

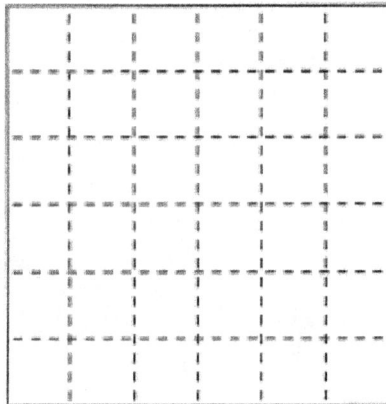 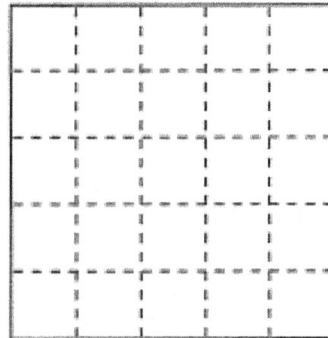

 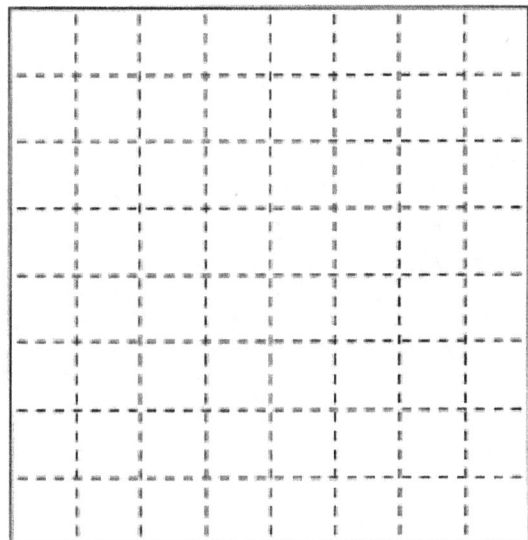

IMP Year 2, Do Bees Built It Best? Unit, Teacher's Guide

© 2010 Interactive Mathematics Program

149

IMP Year 2, Do Bees Built It Best? Unit, Teacher's Guide

© 2010 Interactive Mathematics Program

151

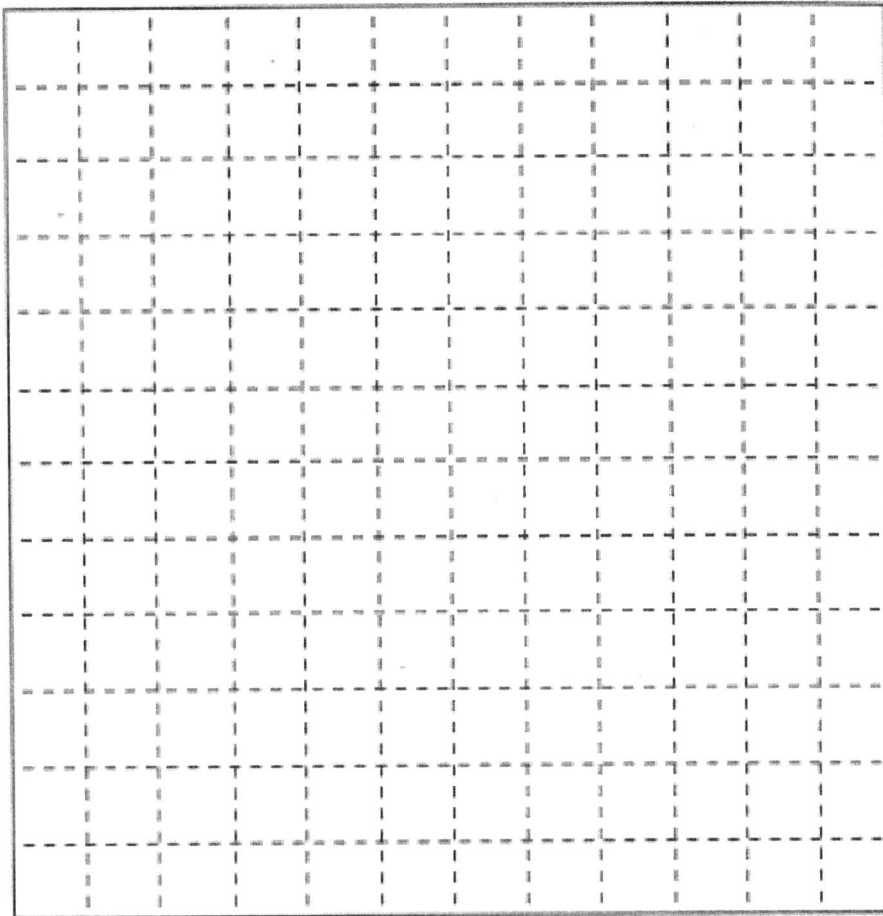

IMP Year 2, Do Bees Built It Best? Unit, Teacher's Guide

© 2010 Interactive Mathematics Program

152

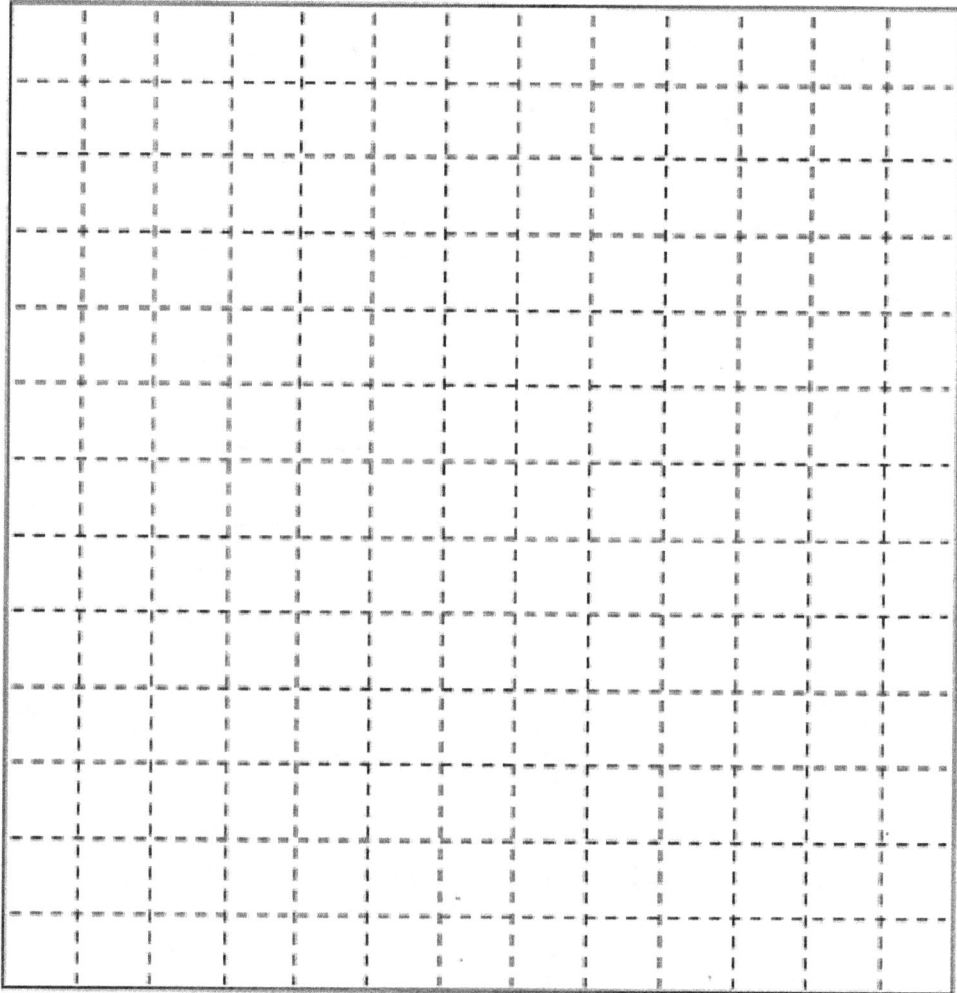

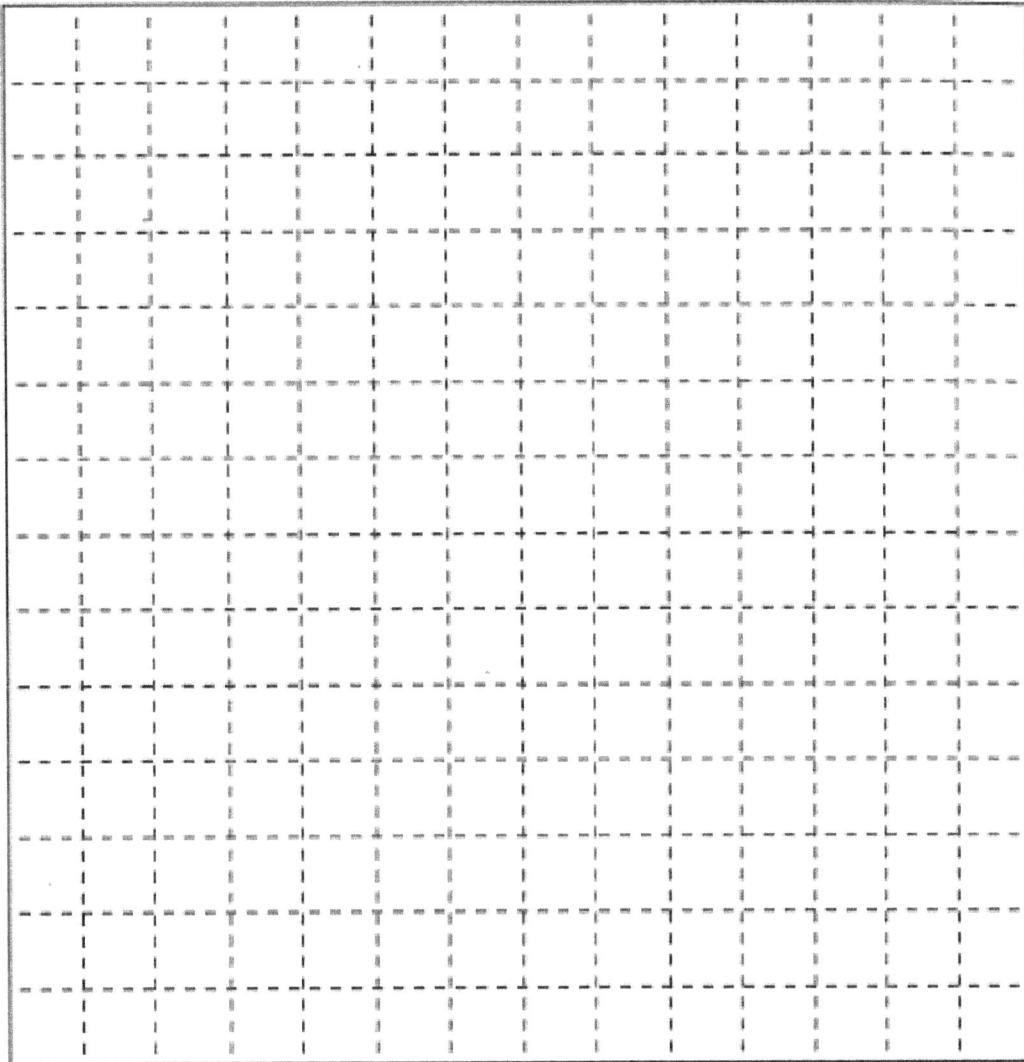

Assessments

In-Class Assessment

A rancher is building a rectangular corral. He is using one wall of his barn as one of the sides.

The barn wall is quite long, so the rancher only needs fencing along the other three sides. He has 300 feet of fencing.

1. The rancher wants the area of the corral to be as large as possible. What should he choose as the dimensions of the corral? (Don't assume that the "obvious" answer is correct.)

2. Use l to represent the length (in feet) of the side of the corral opposite the barn wall. Find a formula that expresses the area of the corral in terms of l.

3. Use your answer to Question 2 and a graphing calculator to justify your answer to Question 1. Write a summary of your reasoning.

Take-Home Assessment

Part I: Heptagonal Swimming

The city of Euclid is building a public swimming pool. The architect of the pool is a geometer at heart. She decides to build the pool in the shape of a prism whose base is a regular heptagon. A heptagon is a seven-sided polygon.
The walls of the pool will be vertical. The pool will be 4 feet deep and will have a perimeter of 350 feet.

1. After the pool is built, the city will paint the inside of the pool. This means painting the seven walls as well as the bottom. Find the total surface area of the inside of the pool.

2. Once the pool is painted, the city will fill it with water. Find the volume of water needed to fill the pool. Assume the pool is filled all the way to the top.

3. If a swimmer swims from one corner of the pool to the center of the opposite side (shown by the arrow), how far will she swim?

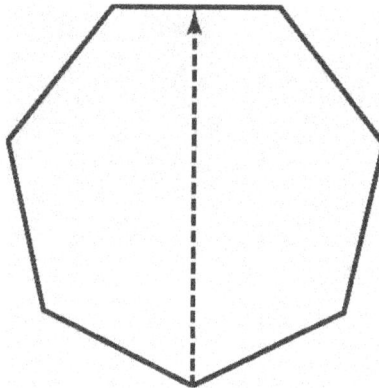

Part II: Are They Equal?

The polygon *ABCD* shown here is a square. Point *E* is the midpoint of $\overline{AB}$. Point *F* is the midpoint of $\overline{BC}$. Point *D* is connected to each of the points *E, B,* and *F*. This divides the square into four triangles: Δ*DAE*, Δ*DEB*, Δ*DBF*, and Δ*DFC*.

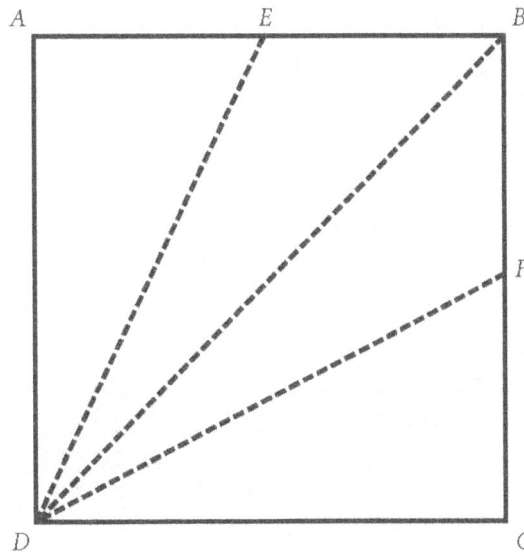

How do the areas of the triangles compare? Show that your conclusions hold true for *any* square.

Do Bees Build It Best? Guide for the TI-83/84 Family of Calculators

This guide gives suggestions for selected activities of the Year 2 unit *Do Bees Build It Best?* The notes that you download contain specific calculator instructions that you might copy for your students. NOTE: If your students have the TI-Nspire handheld, they can attach the TI-84 Plus Keypad (from Texas Instruments) and use the calculator notes for the TI-83/84.

The primary calculator topics in *Do Bees Build It Best?* are the use of the trigonometric functions and their inverses and the use of square roots when applying the Pythagorean theorem. Both topics require attention to order of operations, but both are relatively straightforward. The biggest calculator pitfall is likely to be using radian mode rather than degree mode in trigonometric calculations.

Students will need at least a scientific calculator for several of the homework assignments. If students have not yet purchased one, encourage them to do so at this time. It is important that they become comfortable with using the trigonometric and square-root functions on their own calculators, which may work quite differently from the TI graphing calculator.

Although this unit does not introduce many new calculator applications, it does provide many opportunities to experience how the graphing feature can help students understand what is happening algebraically. *Don't Fence Me In* is a powerful example of this.

The series of corral problems, culminating in the discussion of *Building The Best Fence*, provides a perfect context within which to review many of the calculator skills acquired so far: using the previous entry, graphing, tables, function notation, and programming.

More Gallery Measurements: This assignment provides a good opportunity to discuss the use of trigonometric functions on the TI graphing calculator. Throughout this unit, you may need to remind students that degrees represent only one of several ways of measuring angles, and that their calculators should be set to degree mode rather than radian mode. Radian mode is the calculator's default setting. The calculator will reset to the radian mode whenever the calculator is reset, and frequently also when the batteries are changed. Because many students may be sharing the class set of TI calculators, they should make a habit of checking the mode every time they use the trigonometric functions. To set the calculator properly, press MODE. If necessary, move the cursor to **DEGREE** and press ENTER. Exit the MODE screen by pressing 2nd [QUIT] or CLEAR.

On Question 1, caution students to use care in entering the product of 94 and sin 30°. On paper, sin 30 • 94 is ambiguous. In the calculator, it can lead to improper evaluation. The TI-83/84 automatically inserts a left parenthesis after the $\boxed{\text{SIN}}$ key and assumes closure is at the end of the line if no right parenthesis is entered. As shown here, this will cause it to interpret this key sequence as sin(30° • 94).

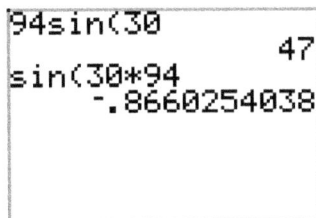

```
94sin(30
               47
sin(30*94
      -.8660254038
```

Sailboats and Shadows: If students do this activity at home, they need to know how to use the trigonometric functions on their personal calculators. Whereas TI graphing calculators require entry of the function followed by the angle, most scientific calculators require entry of the angle first. Also, they will need to be sure that their own calculators are set in degree mode. If students do not yet have their own scientific (or graphing) calculators for home use, remind them that they can still obtain a trigonometric expression for each answer in this activity; then, when they come to class, they can quickly evaluate those expressions using the classroom calculators.

Any Two Sides Work: Remind students that the square-root function is found on the TI calculator by pressing $\boxed{\text{2nd}}$ $\boxed{\sqrt{}}$.

Leslie's Fertile Flowers: After discussing the algebraic solution of this activity, you can have students check the solution graphically. They can enter each side of the equation $13^2 - x^2 = 15^2 - (14 - x)^2$ as a separate function on the $\boxed{\text{Y=}}$ screen and then look for the intersection of the two graphs. You might use **0:ZoomFit** to set the vertical range to fit the function within the horizontal range you have selected.

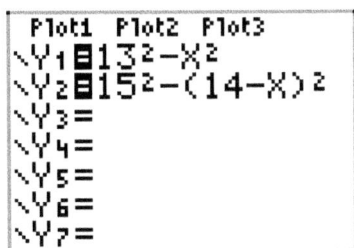

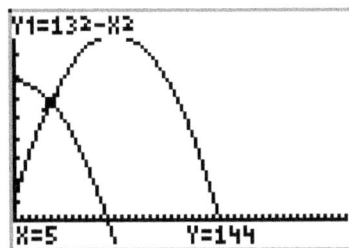

```
Plot1  Plot2  Plot3
\Y1◻13²-X²
\Y2◻15²-(14-X)²
\Y3=
\Y4=
\Y5=
\Y6=
\Y7=
```

```
Y1=13²-X²



X=5          Y=144
```

Don't Fence Me In: If you plan to use the *In-Class Assessment for "Do Bees Build It Best?"* provided in the *Teacher's Guide,* it is worth spending a little extra time discussing the use of a graphing calculator in this activity. If necessary, refer students to the calculator note "Graphing Functions on the Graphing Calculator." If students need more practice later with finding maximum values for an expression by graphing, the supplemental problems *Finding the Best Box* and *Another Best Box* both involve that concept.

More Fencing, Bigger Corrals: If you pursue the optional topic "Area Formulas in Terms of Perimeter" in the *Teacher's Guide,* you will find that the graphs of the resulting area formulas are interesting to examine as well. Students can also trace the functions on the graph to verify that the formulas work for a perimeter of 300.

Simply Square Roots: When introducing this activity, remind students to be careful with the use of parentheses if they use the calculator to compare expressions involving square roots. Also point out that what students are looking for in Question 4 are not decimal approximations from the calculator, but other expressions that are exact answers and that may still involve a radical sign.

Building the Best Fence: The *Teacher's Guide* suggests an extension of the discussion of this activity using the graphing calculator, in the section "Summing up Corrals." Several calculator methods are offered. Depending on the amount of experience your class has with various calculator techniques and the amount of class time you have available, select one or more of these methods to pursue. As an alternative, you might assign a different method to each group and give them time to report and demonstrate to the class. In any case, make sure students get a chance to see what the corral formula looks like graphically. It makes a vivid conclusion to the series of corral problems. Although most of the necessary calculator skills have been covered previously, the calculator note "Summing Up Corrals" provides review in case your students are not yet completely familiar with all of these techniques.

Leslie's Floral Angles: The inverse sine, inverse cosine, and inverse tangent functions are secondary operations on the $\boxed{\text{SIN}}$, $\boxed{\text{COS}}$, and $\boxed{\text{TAN}}$ keys. To find $\sin^{-1} 0.5$, for example, press $\boxed{\text{2nd}}$ $\boxed{\text{SIN}^{-1}}$ $\boxed{.}$ $\boxed{5}$. As needed, remind students to be sure that their calculators are in degree mode, and to use parentheses where necessary.

In-Class Assessment: If you reset the calculators between classes during assessments, be sure to remind students to return the calculator to degree mode.

Supplemental Problems:

Hero and His Formula: Students interested in programming could write a program based on Hero's formula. If some guidance is needed, see the calculator note "Introduction to Programming the Calculator."

Finding the Best Box and *Another Best Box:* The graphing calculator will be useful in finding the maximum value for algebraic expressions in these two activities.

Graphing Functions on the Graphing Calculator

Here is a summary of how to graph functions on a TI graphing calculator.

1. Press $\boxed{\text{Y=}}$ to bring up the window where you will enter your function. To display your graph, you will need to replace your variables with *X* and *Y*. Remember that *Y* will be the vertical axis of your graph and should be the dependent variable. Following **Y1=,** enter your function. In this case, let's consider the function $y = x(10 - x)$, so enter **X(10−X).** Use the arrow and $\boxed{\text{CLEAR}}$ keys as necessary to delete any other functions that were previously entered.

```
Plot1  Plot2  Plot3
\Y1◼X(10-X)
\Y2=
\Y3=
\Y4=
\Y5=
\Y6=
\Y7=
```

2. Press $\boxed{\text{WINDOW}}$ to access the options for adjusting the viewing window. **Xmin, Xmax, Ymin,** and **Ymax** represent the minimum and maximum coordinates that will be displayed on your graph screen. **Xscl** and **Yscl** control the distance between tick marks on your axes and can usually be ignored. Use the arrow keys to move among these variables as you set them to the desired values. The calculator has several features that can help you quickly adjust the viewing window. But there is no substitute for simply taking a moment to ask yourself what range of values makes sense for the situation that your function represents.

```
WINDOW
 Xmin=-10
 Xmax=10
 Xscl=1
 Ymin=-10
 Ymax=10
 Yscl=1
 Xres=1
```

3. Press $\boxed{\text{GRAPH}}$ to view your graph. You may need to return to the Window settings if your viewing window is not satisfactory.

4. If you have created an In-Out table, check to verify that the values from your table are on the graph of the function by pressing $\boxed{\text{TRACE}}$. Coordinates for X and Y should be displayed at the bottom of the screen. Use the left and right arrow keys to move the cursor along the graph of your function. If no coordinates appear, you may have turned them off under the **FORMAT** menu, accessed by pressing $\boxed{\text{2ND}}$ [FORMAT]. To make the coordinates appear, highlight **CoordOn** and press $\boxed{\text{ENTER}}$. In any case, you may want to take a moment to familiarize yourself with the options available in the **FORMAT** menu shown here.

```
RectGC  PolarGC
CoordOn  CoordOff
GridOff  GridOn
AxesOn  AxesOff
LabelOff  LabelOn
ExprOn  ExprOff
```

Adjusting the Viewing Window

While tracing, you may find it difficult to position the cursor on the exact points from your In-Out table. This is because the calculator displays lines by illuminating square dots, called *pixels,* and the number of pixels across the screen may not work well with the viewing window you have chosen. If you need more accuracy, you can zoom in on your point. Or, you can go directly to a chosen value of *X* while in the Trace mode simply by typing the desired value and then pressing $\boxed{\text{ENTER}}$.

IMP Year 2, Do Bees Build It Best? Unit, Calculator Notes

© 2010 Interactive Mathematics Program

162

Zoom In and Zoom Out

Press ZOOM and select **2:Zoom In.** When your graph reappears, it will not yet have changed. Use the arrow keys to position your cursor at the point where you would like the new screen centered. Press ENTER to zoom in on that point. If you return to the WINDOW screen, you will notice that what you have really done is redefine the viewing window. Press TRACE to return to the Trace mode.

You can **Zoom In** repeatedly for as much accuracy as needed or **Zoom Out** to view a larger portion of the graph. Pressing ZOOM and selecting **6: ZStandard** is a quick way to reset the viewing window to –10 to 10 in each direction.

Zoom Box

Because the cursor location defines the center of the screen when you zoom in, it can often be difficult to zoom to a window that still contains everything you want to view. Another ZOOM option, **ZBox,** is especially useful.

Press ZOOM and select **1:ZBox.** Use the arrow keys to move the cursor to the upper-left corner of a rectangle you'd like to zoom in on. Press ENTER. Now move the cursor to the bottom-right corner of the rectangle you'd like to zoom in on. A box will be drawn around the area that you wish to view in more detail. Press ENTER again, and the screen will be redrawn, zooming in on the area that you have boxed.

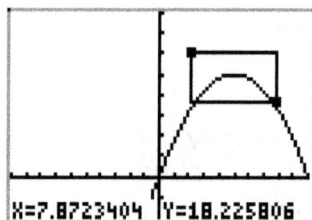

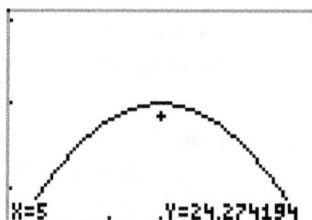

A Friendly Window

ZOOM and WINDOW provide relatively easy ways to adjust the window. But as you noticed previously, the resulting pixel size is probably not convenient if you are trying to trace to a particular value of X. You can tell the calculator exactly what increment of X you would like the width of each pixel to represent. This is controlled by a variable called ΔX (delta X), which is found on the **Window** submenu of the VARS function.

For example, if you want your cursor to move in X-increments of 0.1, press 2ND [QUIT] to return to the home screen. Press . 1 STO> VARS and then ENTER to select the 1:**Window** submenu. Use the down arrow key to scroll down to 8:ΔX, press ENTER again to copy this variable to your home screen, and press ENTER one more time to execute the command.

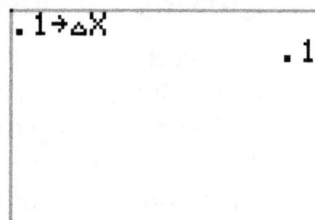

This process will leave **Xmin** unchanged and change **Xmax** to **Xmin + 94 • ΔX.**

IMP Year 2, Do Bees Build It Best? Unit, Calculator Notes

163

© 2010 Interactive Mathematics Program

Press $\boxed{\text{TRACE}}$ to return to your graph, and you will find that the cursor now moves in increments of 0.1 in the X-direction. In a similar manner, you can specify any value to be the X-increment when you trace. For example, if **Xmin** is an integer, then the statement **1→ΔX** would result in only integer values for X.

IMP Year 2, Do Bees Build It Best? Unit, Calculator Notes

164

© 2010 Interactive Mathematics Program

Summing Up Corrals

Your work on *Building the Best Fence* probably resulted in a formula for the area of a regular polygon with *n* sides and a perimeter of 300 similar to the one presented here.

$$A = n \cdot \frac{1}{2} \cdot \frac{300}{n} \cdot \frac{150}{n} \cdot \tan\left(90° - \frac{180°}{n}\right)$$

The calculator provides a variety of ways to use this formula to explore what happens to the area of a regular polygon of fixed perimeter as the number of sides is increased.

Trying Specific Values

1. The most straightforward way to explore what happens as the number of sides changes is simply to evaluate this expression for several values of *n*. Before you do this, press MODE and make sure that your calculator is in the **DEGREE** mode. Be careful with order of operations. You may want to use several sets of parentheses to avoid confusion.

2. After you have entered this expression several times for different values of *n*, you might suspect that there must be a better way. There is! Two methods are described here.

   a. After you have entered and evaluated the expression with one value of *n*, press 2ND [ENTRY]. The expression will be recalled to the screen, as shown here, and you can use the arrow keys to move the cursor around for editing. Simply replace the previous value of *n* with a new number and press ENTER. Don't forget to replace the value of *n* in all four places it appears.

   ```
   3*(1/2)*(300/3)*
   (150/3)*tan(90-1
   80/3)
            4330.127019
   3*(1/2)*(300/3)*
   (150/3)*tan(90-1
   80/3)■
   ```

   If you are changing to a value of *n* that has a larger number of digits, you will have to insert space for the extra digits in the expression. Move the cursor to the character that will follow the inserted character. Press 2ND [INS] to shift into the Insert mode. The cursor will switch from a flashing block to an underline. Enter the character to be inserted.

   b. Here's another method. As shown here, first enter your value of *n* and tell the calculator to store that as *N* (using the STO> key), but do not press ENTER yet. On the same line, enter a colon by pressing ALPHA [.], then enter the area expression. In this expression, however, use *N* instead of a specific value of *n*. Press ENTER to evaluate the expression for the chosen value of *n*. For the next value, press 2ND [ENTRY] to recall the expression. Now all you have to change is the single number that you stored as *N*.

   ```
   3→N:N*(1/2)*(300
   /N)*(150/N)*tan(
   90-180/N)
            4330.127019
   ```

IMP Year 2, Do Bees Build It Best? Unit, Calculator Notes

© 2010 Interactive Mathematics Program

165

Graphing

Graphing gives a vivid picture of what happens with the area as the number of sides increases.

1. Enter the area expression as a function after pressing $\boxed{\text{Y=}}$. Remember that you must use *X* instead of *n*.

2. Press $\boxed{\text{WINDOW}}$ and set the range variables to reasonable values for this problem. When setting **Xmin,** consider the smallest number of sides a polygon can have. Choose some reasonable number for **Xmax.** Your experience with the corrals should have given you a good feel for the likely range of values for **Ymin** and **Ymax.** Think about what that variable represents.

```
Plot1 Plot2 Plot3
\Y1■X(1/2)(300/X
)(150/X)tan(90-1
80/X)
\Y2=
\Y3=
\Y4=
\Y5=
```

3. Because the number of sides must be an integer, you may want to set the *X*-increment interval to 1 for tracing. (*Note:* This will likely change **Xmax.**) To do so, press $\boxed{\text{2ND}}$ [QUIT] to return to the home screen. Enter $\boxed{1}$ $\boxed{\text{STO>}}$ $\boxed{\text{VARS}}$. Press $\boxed{\text{ENTER}}$ to select **WINDOW,** scroll down to 8:Δ**X,** and press $\boxed{\text{ENTER}}$ twice more.

```
1→ΔX
                    1
```

4. Now press $\boxed{\text{TRACE}}$ and use the right and left arrow keys to observe what happens to the area as the number of sides increases. How many sides must the polygon have before adding another side seems to make little difference? What seems to be the limit to the area of a regular polygon with a perimeter of 300?

5. Compare that apparent limit to the area of a circle with a circumference of 300. (The perimeter of a circle is called the *circumference.*) After all, if you add enough sides, the polygon begins to look very much like a circle. The circumference of a circle is found by the formula $C = 2\pi r$, where *r* is the radius of the circle. The area of a circle is defined by $A = \pi r^2$. (π is the Greek character *pi* and is the $\boxed{\text{2ND}}$ function of the $\boxed{\wedge}$ key.)

6. Determine whether you can combine these formulas to find the area of a circle with a circumference of 300. How does this compare to the area of the polygon with a large number of sides, as shown on your graph?

Using a Table

You can also use the area function to create an In-Out table in the calculator, with the number of sides of the polygon as the *In* and the area as the *Out.*

1. Enter the area expression at $\boxed{\text{Y=}}$ as described in the previous section, *Graphing.*

2. Set up the table by pressing $\boxed{\text{2ND}}$ [TBLSET]. Because you may want to look at a wide range of numbers of sides for

```
TABLE SETUP
 TblStart=
 ΔTbl=1
Indpnt: Auto Ask
Depend: Auto Ask
```

your polygons, tell the calculator to generate table values only for the input values that you specify. Ignore **TblStart** and **ΔTbl** and set **Indpnt:** to **Ask** and **Depend:** to **Auto.** (**TblStart** tells the calculator where to start the *In* values for the table if **Indpnt:** is set to **Auto**, and **ΔTbl** tells it how far apart each input value should be.)

3. Press 2ND [TABLE] to view the table. Enter a value for the number of sides and press ENTER. The calculator will automatically enter the area of the polygon on the output side of the table.

X	Y₁	
3	4330.1	

X=

Using Function Notation

The calculator will also let you evaluate the function through the use of function notation.

1. Enter the area expression at Y= as described in the section *Graphing*.

2. Return to the home screen by pressing 2ND [QUIT]. Press VARS and use the right arrow key to highlight the **Y-VARS** menu heading. Press ENTER to select 1:**Function** and ENTER once more to select **Y1.**

3. When **Y1** appears on the home screen, press ((and enter the number of sides for which you would like the function evaluated. Then press)) ENTER. To repeat with another value, press 2ND [ENTRY] and edit the previous entry.

Y₁(3)
4330.127019

Programming

Try to write a program that will request the number of sides as an input and display the area of the polygon as an output.

Introduction to Programming the Calculator

Starting the Program

If you press PRGM, you will see three menu headings across the top of the screen: **EXEC,** from which to execute (run) a program; **EDIT,** from which to edit (change) an existing program; and **NEW,** from which to begin entering a new program.

To start a new program, press PRGM, use the right arrow key to highlight the **NEW** menu heading, and press ENTER to select **1:Create New.**

```
EXEC EDIT NEW
1:Create New
```

Notice that the cursor is now a flashing **A.** This means that the calculator is locked in the **ALPHA** mode. Enter a program name of eight characters or fewer and press ENTER.

Entering a Program

Every program consists of a series of commands that the calculator executes in order. The TI calculator marks the beginning of each command with a colon punctuation mark, which the calculator inserts when you press ENTER at the end of the previous line. You can also enter more than one command on a single line by separating each command with a colon.

```
PROGRAM:POLYGON
:Input P
:Disp "NUMBER OF
 SIDES?"
:Input N
:P²/(4N)→B
:tan(90-180/N)→C
```

You enter most commands into your program by selecting them from menus that are displayed when various keys are pressed. You cannot enter these commands by spelling them out in ALPHA mode.

Occasionally, you will accidentally select the wrong menu when looking for a program command to insert into your program. When that happens, press CLEAR to return to your program editing. (If you press 2ND [QUIT], the calculator will exit the program editing mode altogether. You will then have to select your partially completed program for editing, which will be described later.)

Running a Program

When you finish entering a program, exit the editing mode by pressing 2ND [QUIT].

Select a program to run by pressing PRGM, highlighting the name of the program in the **EXEC** menu, and then pressing ENTER. The program name will appear on the home screen. Press ENTER again to run the program.

If you need to interrupt your program while it is running, press ON.

IMP Year 2, Do Bees Build It Best? Unit, Calculator Notes

© 2010 Interactive Mathematics Program

168

Editing a Program

Don't be surprised if your program does not operate as intended the first time. You will usually have to return to the editing mode to revise a new program. These glitches in the program are known as *bugs*, and the process of correcting them is known as *debugging.*

To edit your program, press PRGM, highlight the **EDIT** menu heading, highlight the name of your program, and press ENTER.

Programming Commands

The essence of programming is stringing together a series of commands into a logical sequence that will accomplish something you intend. The TI calculator has a large number of commands available that are fully described in the calculator's manual. Among these are a few basic commands that will enable you to accomplish a wide range of tasks; these commands are described here. Similar commands exist in many programming languages.

Input/Output Commands

Start a new program, giving it a title, so that you will be able to try these commands. Some of the keys, such as PRGM, cause different menus to be displayed when you are editing a program than when you are at the home screen.

```
CTL I/O EXEC
1:Input
2:Prompt
3:Disp
4:DispGraph
5:DispTable
6:Output(
7↓getKey
```

From within a new program, press PRGM and use the right arrow key to highlight the **I/O** menu heading. The commands found in this menu control input and output of information. We will cover a few of the most important of these first, because any program is useless unless it can give you information.

When your program is executed, an **Input** command will cause the program to display a question mark and to pause until you enter a value. In your program, **Input** must be followed on the same line by the name of the variable to which the value is to be saved. Note the example shown here.

```
PROGRAM:INOUT
:Input A
:Disp "AND FIVE
IS"
:Disp A+5
:
```

The **Disp** (display) command is followed by the number, variable, expression, or string of text that is to be displayed. Text to be displayed must be enclosed in quotation marks. **8:ClrHome** clears the calculator screen.
The example program shown here asks for a value of *A* and then displays some text followed by the result of adding *A* + 5.

```
prgmINOUT
?7
AND FIVE IS
            12
         Done
```

IMP Year 2, Do Bees Build It Best? Unit, Calculator Notes

© 2010 Interactive Mathematics Program

169

Variables

The program must store all intermediate answers as variables. Variable names must be single letters. When you do a computation that uses the value of a variable, you can store the answer either as a new variable or as a new value for the same variable. For example, the program shown here would set A equal to 5, multiply A by 2 and store the answer as B, and then subtract 3 from B and store the answer as a new value of B.

```
PROGRAM:VBLES
:5→A
:2A→B
:B-3→B
```

Branching

Branching commands change the order in which the calculator executes the commands in a program. You enter branching commands by pressing PRGM and then selecting the command from the **CTL** (control) menu.

```
PROGRAM:BRANCH
:5→A
:Goto B
:A+7→A
:Lbl B
:Disp A
```

The simplest branching command is **Goto,** which must be accompanied by a **Lbl** (label) command. You must scroll down the menu to find both of these commands. The **Lbl** command must be followed by a letter or number. This command simply labels that line in the program with that letter or number. The **Goto** command must then be followed by the name of the labeled place to which you wish the program to go next. In the program BRANCH, shown here, the third line is never executed, because the **Goto** statement causes it to be skipped.

```
PROGRAM:IF
:Input X
:If X≥10
:Disp "BIG"
:Disp X
```

One of the most powerful branching statements is the **If** statement, also found on the PRGM **CTL** menu. The **If** statement must always be accompanied by a test of some kind, such as **A>3.** If the test statement is true, then the program goes on to execute the next line. If it is not, the program skips the next line. The comparison symbols, such as = and ≥, are found by pressing 2ND [TEST].

For example, the program shown above asks the user to supply a value of X. If the number entered is at least 10, the calculator will display the word "**BIG,**" followed by the number. If the number is less than 10, the program skips the command to display the word "**BIG**" but still displays the number. These next two screens show possible results of running this program.

```
prgmIF
?18
BIG
            18
         Done
```

```
prgmIF
?3
             3
          Done
```

If you want more than one command to be executed when the **If** statement is true, enclose those commands between **Then** and **End,** also found on the PRGM **CTL**

IMP Year 2, Do Bees Build It Best? Unit, Calculator Notes

© 2010 Interactive Mathematics Program

170

menu. The program then says, "**If** this statement is true, **Then** do all of the things listed below until you find **End.** If it is not true, skip everything from here to **End.**"

The next program asks the user to enter the number 5. When a number is entered, the program assigns the number to the variable X and checks to confirm whether it is 5. If it is 5, then the program displays "**THANK YOU**" and stops. If it is not 5, the program skips to the **Goto** statement, which tells it to go back to the beginning and ask again for 5 to be entered. (**Stop** is found on the $\boxed{\text{PRGM}}$ **CTL** menu also and is discussed further in *More Control Commands.*)

PROGRAM: STOP
:Lbl A
:Disp "ENTER 5"
:Input X
:If X=5
:Then
:Disp "THANK YOU"
:Stop
:End
:Goto A

```
prgmSTOP
ENTER 5
?7
ENTER 5
?5
THANK YOU
            Done
```

The screen shown here illustrates how the program reacts if the user enters 7 or 5.

Loops

When you want a program to do something several times, it is often helpful to use a **For** command to create a loop. 4:**For** is also on the $\boxed{\text{PRGM}}$ **CTL** menu. The **For** command includes a set of parentheses with four entries separated by commas. These four entries, in order, are the name of a variable, the starting and ending values of that variable, and the increment (the number you will count by to get from the starting value to the ending value).

When the program executes the **For** command, it begins by setting the named variable equal to the starting value. It then continues to the following lines until it encounters an **End** statement. At the **End** statement, the calculator adds the increment to the variable and tests whether the variable has exceeded the ending value. If not, the program loops back to the **For** statement and starts through the loop again. If the variable has exceeded the ending value, the program exits the loop by going to the command that follows **End.**

This is most easily understood through an example. The commands shown here will cause the program to count from 12 to 18 by 2's and then display "**BYE.**"

```
PROGRAM:LOOP
:For(N,12,18,2)
:Disp N
:End
:Disp "BYE"
:Stop
```

```
prgmLOOP
            12
            14
            16
            18
BYE
       Done
```

More Control Commands

Two more commands on the PRGM **CTL** menu are especially useful. The **Stop** command tells the program to quit right there. If there is no **Stop** statement, the program assumes that it is to stop after executing the last line (unless that line explicitly sends the program to another line). Always include a way for the program to end. Avoid endless loops that would cause the program to keep running forever.

The **Pause** command tells the program to stop so that you can read what is displayed on the screen. If the program displays many things, some may scroll off the screen too quickly for you to read them. When the **Pause** command is executed, the program stops until the ENTER key is pressed, and then it resumes.

Practice Exercises

The best way to learn all of this is to write some simple programs that use these commands and find out how they work. Here are a few problems to get you started. Find out if you can write programs to accomplish each of these tasks.

1. Display your name five times.

2. Ask for a number to be entered, and then display "**POSITIVE**" or "**NEGATIVE**" or "**ZERO**," depending on the value of the number.

3. Display a countdown from 10 to 0.
    There are many more programming commands available on the graphing calculator. As you get comfortable with the commands described here, explore the calculator's manual and note what else you can find.

www.ingramcontent.com/pod-product-compliance
Lightning Source LLC
Chambersburg PA
CBHW051345200326
41521CB00014B/2486